Human Pluripotent Stem Cells

Human Pluripotent Stem Cells

A Practical Guide

Uma Lakshmipathy
Chad C. MacArthur
Mahalakshmi Sridharan
Rene H. Quintanilla

Cell Biology, Life Sciences Solutions,
Thermo Fisher Scientific, USA

Registered Offices
John Wiley & Sons, Inc., 111 River Street, Hoboken, NJ 07030, USA

Editorial Office
111 River Street, Hoboken, NJ 07030, USA

For details of our global editorial offices, customer services, and more information about Wiley products visit us at www.wiley.com.

Wiley also publishes its books in a variety of electronic formats and by print-on-demand. Some content that appears in standard print versions of this book may not be available in other formats.

Library of Congress Cataloguing-in-Publication Data

Names: Lakshmipathy, Uma, author. | MacArthur, Chad C., author. | Sridharan,
    Mahalakshmi, author. | Quintanilla, Rene H., author.
Title: Human pluripotent stem cells : a practical guide / Uma Lakshmipathy,
    Chad C. MacArthur, Mahalakshmi Sridharan, Rene H. Quintanilla.
Description: Hoboken, NJ : John Wiley & Sons, 2018. |
    Includes bibliographical references and index. |
Identifiers: LCCN 2017036447 (print) | LCCN 2017046356 (ebook) |
    ISBN 9781119394341 (pdf) | ISBN 9781119394365 (epub) |
    ISBN 9781119394334 (cloth)
Subjects: LCSH: Multipotent stem cells. | Stem cells.
Classification: LCC QH588.S83 (ebook) | LCC QH588.S83 L35 2018 (print) |
    DDC 616.02/774–dc23
LC record available at https://lccn.loc.gov/2017036447

Cover Design: Wiley
Cover Image: Courtesy of Rene H. Quintanilla Jr.

Set in 10/12pt Warnock by SPi Global, Pondicherry, India
Printed in Singapore by C.O.S. Printers Pte Ltd

10 9 8 7 6 5 4 3 2 1

Contents

1

Introduction

Successful execution of any cell-based project relies on a setting up a robust cell culture laboratory. Guidelines under the Guidance on Good Cell Culture Practice provides an overview of the critical parameters in establishing facilities and training personnel [1]. This is even more important for a stem cell laboratory where primary cells from donor tissue or their derivatives are cultured for extended periods of time [2].

1.1 Biosafety

In addition to the safety risks common to most workplaces, such as electrical and fire hazards, a cell culture laboratory has a number of specific hazards associated with handling and manipulating human or animal cells and tissues, as well as toxic, corrosive or mutagenic solvents and reagents. The most common of these hazards are accidental inoculations with syringe needles or other contaminated sharps, spills and splashes onto skin and mucous membranes, ingestion through mouth pipetting, animal bites and scratches, and inhalation exposures to infectious aerosols.

The fundamental objective of any biosafety program is to reduce or eliminate exposure of laboratory workers and the outside environment to potentially harmful biological agents. The most important element of safety in a cell culture laboratory is strict adherence to standard microbiological practices and techniques.

Human Pluripotent Stem Cells: A Practical Guide, First Edition. Uma Lakshmipathy, Chad C. MacArthur, Mahalakshmi Sridharan and Rene H. Quintanilla.

1.2 Biosafety Levels

The regulations and recommendations for biosafety in the United States are contained in the document *Biosafety in Microbiological and Biomedical Laboratories*, prepared by the Centers for Disease Control (CDC) and the National Institutes of Health (NIH), and published by the US Department of Health and Human Services. The document defines four ascending levels of containment, referred to as biosafety levels 1 through 4, and describes the microbiological practices, safety equipment, and facility safeguards for the corresponding level of risk associated with handling a particular agent.

- *Biosafety Level 1 (BSL-1)*: the basic level of protection common to most research and clinical laboratories. Appropriate for agents that are not known to consistently cause disease in normal, healthy human adults (examples: *Bacillus subtilis, E. coli*).
- *Biosafety Level 2 (BSL-2)*: appropriate for moderate-risk agents known to cause human disease of varying severity by ingestion or through percutaneous or mucous membrane exposure. Most cell culture labs should be at least BSL-2, and all stem cell labs have this as a requirement.
- *Biosafety Level 3 (BSL-3)*: BSL-3 is appropriate for indigenous or exotic agents with a known potential for aerosol transmission, and for agents that may cause serious and potentially lethal infections.
- *Biosafety Level 4 (BSL-4)*: BSL-4 is appropriate for exotic agents that pose a high individual risk of life-threatening disease by infectious aerosols and for which no treatment is available. These agents are restricted to high-containment laboratories.

For more information about the biosafety level guidelines, refer to *Biosafety in Microbiological and Biomedical Laboratories*, 5th edition, which is available for downloading at www.cdc.gov/biosafety/.

1.3 Aseptic Technique

Successful cell culture depends heavily on keeping the cells free from contamination by microorganisms such as bacteria, fungi, and viruses. Non-sterile supplies, media, and reagents, airborne particles laden with microorganisms, unclean incubators, and dirty work surfaces are all sources of biological contamination.

Aseptic technique, designed to provide a barrier between the microorganisms in the environment and the sterile cell culture, depends upon a set of procedures to reduce the probability of contamination from these sources. The elements of aseptic technique are a sterile work area, good personal hygiene, sterile reagents and media, and sterile handling.

1.3.1 Maintaining a Sterile Work Area

The simplest and most economical way to reduce contamination from airborne particles and aerosols (e.g., dust, spores, shed skin, sneezing) is to use a cell culture hood.

- The cell culture hood should be properly set up, and located in an area that is restricted to cell culture, is free from drafts from doors, windows, and other equipment, and with no through traffic.
- The work surface should be uncluttered and contain only items required for a particular procedure; it should not be used as a storage area.
- Before and after use, the work surface should be disinfected thoroughly, and the surrounding areas and equipment should be cleaned routinely.
- For routine cleaning, wipe the work surface with 70% ethanol before and during work, especially after any spillage.
- Using a Bunsen burner for flaming is not necessary or recommended in a cell culture hood.
- Leave the cell culture hood running at all times, turning it off only when it will not be used for extended periods of time.
- Practice good personal hygiene. Wash your hands before and after working with cell cultures. In addition to protecting you from hazardous materials, wearing personal protective equipment also reduces the probability of contamination from shed skin as well as dirt and dust from your clothes.

1.3.2 Aseptic Work Area

The major requirement of a cell culture laboratory is to maintain an aseptic work area that is restricted to cell culture work. Although a separate tissue culture room is preferred, a designated cell culture area within a larger laboratory can be used for sterile handling, incubation, and storage of cell cultures, reagents, and media. The simplest and most economical way to provide aseptic conditions is to use a cell culture hood (i.e., biosafety cabinet).

1.3.3   Cell Culture Hood

The cell culture hood provides an aseptic work area while allowing the containment of infectious splashes or aerosols generated by many microbiological procedures. Three kinds of cell culture hoods, designated as Class II, III, and I, have been developed to meet varying research and clinical needs.

- *Class I* cell culture hoods offer significant levels of protection to laboratory personnel and to the environment when used with good microbiological techniques, but they do not provide cultures with protection from contamination. They are similar in design and airflow characteristics to chemical fume hoods.
- *Class II* cell culture hoods are designed for work involving BSL-1, -2, and -3 materials, and also provide an aseptic environment necessary for cell culture experiments. A Class II biosafety cabinet should be used for handling potentially hazardous materials (e.g., primate-derived cultures, virally infected cultures, radioisotopes, carcinogenic or toxic reagents).
- *Class III* biosafety cabinets are gas tight, and provide the highest attainable level of protection to personnel and the environment. A Class III biosafety cabinet is required for work involving known human pathogens and other BSL-4 materials.

A cell culture hood should be large enough to be used by one person at a time, be easily cleanable inside and outside, have adequate lighting, and be comfortable to use without requiring awkward positions. Keep the workspace in the cell culture hood clean and uncluttered, and keep everything in direct line of sight. Disinfect each item placed in the cell culture hood by spraying it with 70% ethanol and wiping clean.

The arrangement of items within the cell culture hood usually adheres to the following right-handed convention.

- A wide, clear workspace in the center with your cell culture vessels.
- Pipettor in the front right and glass pipettes in the left, where they can be reached easily.
- Reagents and media in the rear right to allow easy pipetting.
- Small container in the rear middle to hold liquid waste.

1.3.3.1 Airflow Characteristics of Cell Culture Hoods

Cell culture hoods protect the working environment from dust and other airborne contaminants by maintaining a constant, unidirectional flow of HEPA-filtered air over the work area. The flow can be horizontal, blowing parallel to the work surface, or vertical, blowing from the top of the cabinet onto the work surface.

Depending on its design, a horizontal flow hood provides protection to the culture (if the air is flowing towards the user) or to the user (if the air is drawn in through the front of the cabinet by negative air pressure inside). Vertical flow hoods, on the other hand, provide significant protection to both the user and the cell culture.

1.3.3.2 Clean Benches

Horizontal laminar flow or vertical laminar flow "clean benches" are not biosafety cabinets; these pieces of equipment discharge HEPA-filtered air from the back of the cabinet across the work surface toward the user, and may expose the user to potentially hazardous materials. These devices only provide product protection. Clean benches can be used for certain clean activities, such as the dust-free assembly of sterile equipment or electronic devices, and they should never be used when handling cell culture materials or drug formulations, or when manipulating potentially infectious materials.

1.3.4 Incubator

The purpose of the incubator is to provide the appropriate environment for cell growth. The incubator should be large enough, have forced air circulation, and should have temperature control to within ±0.2 °C. Stainless steel incubators allow easy cleaning and provide corrosion protection, especially if humid air is required for incubation. Although the requirement for aseptic conditions in a cell culture incubator is not as stringent as that in a cell culture hood, frequent cleaning of the incubator is essential to avoid contamination of cell cultures.

1.3.4.1 Types of Incubators

There are two basic types of incubators, dry incubators and humid CO_2 incubators. Dry incubators are more economical, but require the cell cultures to be incubated in sealed flasks to prevent evaporation. Placing a water dish in a dry incubator can provide some humidity, but this does not allow precise control of atmospheric conditions in the incubator. Humid CO_2 incubators are more expensive, but allow superior control of culture conditions. They can be used to incubate cells cultured in Petri dishes or multiwell plates, which require a controlled atmosphere of high humidity and increased CO_2 tension.

1.4 Storage

A cell culture laboratory should have storage areas for liquids such as media and reagents, for chemicals such as drugs and antibiotics, for consumables such as disposable pipettes, culture vessels, and gloves, for glassware such as media bottles and glass pipettes, for specialized equipment, and for tissues and cells.

Glassware, plastics, and specialized equipment can be stored at ambient temperature on shelves and in drawers; however, it is important to store all media, reagents, and chemicals according to the instructions on the label.

Some media, reagents, and chemicals are sensitive to light; while their normal laboratory use under lighted conditions is tolerated, they should be stored in the dark or wrapped in aluminum foil when not in use.

1.4.1 Refrigerators

For small cell culture laboratories, a domestic refrigerator (preferably one without an autodefrost freezer) is an adequate and inexpensive piece of equipment for storing reagents and media at 2–8 °C. For larger laboratories, a cold room restricted to cell culture is more appropriate. Make sure that the refrigerator or cold room is cleaned regularly to avoid contamination.

1.4.2 Freezers

Most cell culture reagents can be stored at −5 °C to −20 °C; therefore, an ultra-deep freezer (i.e., a −80 °C freezer) is optional for storing most reagents. A domestic freezer is a cheaper alternative to a laboratory freezer. While most reagents can withstand temperature oscillations in an autodefrost (i.e., self-thawing) freezer, some reagents such as antibiotics and enzymes should be stored in a freezer that does not autodefrost.

1.4.3 Cryogenic Storage

Cell lines in continuous culture are likely to suffer from genetic instability as their passage number increases; therefore, it is essential to prepare working stocks of the cells and preserve them in cryogenic storage. Do not store cells in −20 °C or −80 °C freezers, because their viability decreases when they are stored at these temperatures.

There are two main types of liquid nitrogen storage systems, vapor phase and liquid phase, which come as wide-necked or narrow-necked storage containers. Vapor-phase systems minimize the risk of explosion with cryostorage tubes, and are required for storing biohazardous materials, while liquid-phase systems usually have longer static holding times and are therefore more economical.

1.5 Contamination

1.5.1 Biological Contamination

Contamination of cell cultures is easily the most common problem encountered in cell culture laboratories, sometimes with very serious consequences. Cell culture contaminants can be divided into two main categories: chemical contaminants, such as impurities in media, sera, and water, including endotoxins, plasticizers, and detergents, and biological contaminants, such as bacteria, molds, yeasts, viruses, mycoplasma, as well as cross-contamination by other cell lines. While it is impossible to eliminate contamination entirely, it is possible to reduce its frequency and seriousness by gaining a thorough understanding of sources and following good aseptic technique.

1.5.2 Cross-Contamination

While not as common as microbial contamination, extensive cross-contamination of many cell lines with HeLa and other fast-growing cell lines is a clearly established problem with serious consequences. Obtaining cell lines from reputable cell banks, periodically checking the characteristics of the cell lines, and practicing good aseptic technique will help you avoid cross-contamination. DNA fingerprinting, karyotype analysis, and isotype analysis can confirm the presence or absence of cross-contamination in your cell cultures.

1.5.3 Using Antibiotics

Antibiotics should never be used routinely in cell culture, because their continuous use encourages the development of antibiotic-resistant strains and allows low-level contamination to persist, which can develop into full-scale contamination once the antibiotic is removed from media, and may hide mycoplasma infections and other cryptic contaminants. Further, some antibiotics might cross-react with the cells and interfere with the cellular processes under investigation.

Antibiotics should only be used as a last resort and only for short-term applications, and they should be removed from the culture as soon as possible. If they are used in the long term, antibiotic-free cultures should be maintained in parallel as a control for cryptic infections.

1.6 Pluripotent Stem Cells

Pluripotent stem cells (PSC) can divide indefinitely, self-renew and can differentiate and functionally reconstitute into almost any cell in the normal developmental pathway, given the right conditions. There are several kinds of pluripotent stem cells.

- *Embryonic stem cells*: isolated from the inner cell mass of the blastocyst stage of a developing embryo. These early cells were destined to create a fetus following implantation. They can differentiate into the three normal developmental germ layers: ectoderm, mesoderm, and endoderm.
- *Embryonic germ cells*: derived from aborted fetuses. These early cells were destined to become sperm and eggs. They have a totipotent capacity for all three germ layers, as well as the capacity to recreate a fetal state.
- *Embryonic carcinoma cells*: isolated from certain types of tumors seen in fetuses. These cells can be nullipotent or multipotent, but do not have the potential, under normal circumstances, to recapitulate the three germ layers of development.
- *Induced pluripotent stem cells*: generated via ectopic expression of one or more genes, and in some cases through chemically induced methods, to reprogram an adult somatic cell into an embryonic stem cell-like state. They have the capacity for self-renewal and the potential to differentiate into the three developmental germ layers.

Pluripotent stem cells were traditionally maintained on a layer of feeder cells (usually either murine or human embryonic inactivated fibroblasts). Cells can be maintained under these conditions for several passages without compromising their proliferation or self-renewal and differentiation potential. Alternatively, cells can be maintained in feeder-free conditions using specialized media systems on a matrix-coated tissue culture surface. This book will demonstrate the feeder-dependent and feeder-free culture of hESC and hiPSC, which will be referred to in all protocols as PSCs. It will also address the methods commonly used to determine pluripotency, as defined by self-renewal marker expression and differentiation potential.

1.7 Procedures

1.7.1 Helpful Tips and Tricks

- General maintenance of PSC cultures requires daily removal of spent media and replenishment of fresh hPSC media. It is crucial to add fresh bFGF (in feeder-dependent cultures), aseptically, on a daily basis to the prewarmed media prior to adding to the cells. Daily visual inspection of cell morphology is highly recommended to ensure proper growth and for the removal of any differentiating cells or colonies via manual dissection.

- As daily maintenance of the PSC cultures is required, it is helpful to develop an optimal working schedule. PSCs should be split every 3–4 days, based on size and distribution of colonies, generally measured in terms of confluency. Do not allow colonies to overgrow and touch each other as they may begin to spontaneously differentiate. Do not triturate your colonies too hard when passaging them, as overtrituration will lead to single cells and small colony clusters, which tend to spontaneously differentiate. Do not passage your culture in large colony clusters as they will also differentiate, due to high density in the interior of the colonies. Hands-on experience and a keen eye are most important in PSC culture.
- Generally, a manageable schedule is employed as highlighted in Table 1.1.
- If cells cannot be fed both weekend days, you may skip a single day and just feed your cultures an additional half volume (1.5 times the normal volume) of media the day before skipping a day's feeding.
- *Note*: it is acceptable to add additional media and skip a day's feeding when the cells have recently been split; do not attempt during logarithmic growth, as metabolism of the media and cytokines may lead to differentiation and loss of cell health and pluripotency.
- Alternatively, if using Essential 8 medium for culture, you may employ the Essential 8 Flex medium and split guide to allow for weekend-free maintenance of cultures.
- A microscope with a 4× or 5× objective and phase contrast or DIC is ideal to observe PSC morphology (Figure 1.1). Brightfield images at 50× magnification and 100× magnification are difficult to observe (see Figure 1.1a, left panel). Phase contrast images at 50× magnification and 100× magnification are preferable to be able to determine morphology (see Figure 1.1a, right panel). Feeder-dependent PSCs demonstrate tight colony formations surrounded by the supporting feeder cells that typically display fibroblast morphology (see Figure 1.1b).

Table 1.1 Overview of pluripotent stem cell (PSC) culture manipulations per week.

	Feeder-dependent PSC	Feeder-free PSC
Sunday	Skip media change	Skip media change
Monday	1) Prepare iMEF for Tuesday 2) Change media and observe cells	Change media and observe cells
Tuesday	Passage cells (1:4–1:5)	1) Prepare matrix-coated plates 2) Passage cells (1:5–1:8)
Wednesday	Change media and observe cells	Change media and observe cells
Thursday	1) Prepare iMEF for Friday 2) Change media and observe cells	Change media and observe cells
Friday	Passage cells (1:4–1:5)	1) Prepare matrix-coated plates 2) Passage cells (1:5–1:8)
Saturday	Change media with 1.5× volume and observe cells	Change media with 1.5× volume and observe cells

iMEF, inactivated mouse embryonic fibroblasts.

PSCs grow in a compact colony formation with very well-defined borders. They have a high nucleus-to-cytoplasm ratio and the colonies grow in a three-dimensional radial pattern (see Figure 1.1c).

- With differentiation, the typical morphology of PSC is altered. PSC colonies that are beginning to differentiate show a loss of defined edges, where the cells become heterogeneous and less compact. The central core remains compact and it is possible to rescue this colony by scraping out the differentiated cells at the edges (Figure 1.2a). As colonies continue to differentiate, there is further loss of defined edges, emergence of large differentiated cells and a heterogeneous central core that is not typical of undifferentiated pluripotent stem cells (see Figure 1.2b).

(a)

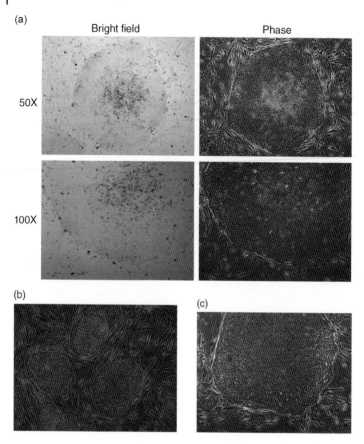

Figure 1.1 **Morphology of intact PSC.** A microscope with a 4× or 5× objective and phase contrast or DIC is ideal to observe PSC morphology. Bright-field images at 50× magnification and 100× magnification are difficult to observe. Phase contrast images at 50× magnification and 100× magnification are preferable to be able to determine morphology (a). 50× magnified phase contrast image of PSC on mitotically inactivated MEF feeders (b). 100× magnified phase contrast image of feeder-dependent PSCs (c).

1.7.2 Daily Monitoring of Morphology and Colony Size

A typical passaging schedule for PSC is given in Table 1.1.

It is important to add fresh media every day, and this includes weekends and holidays (with the previously described exception of skipping one feeding per week). Most often, differentiated

(a)

(b)

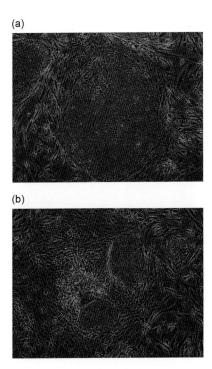

Figure 1.2 Morphology of differentiating PSC. 100× magnified phase contrast image of feeder-dependent PSC with early signs of differentiation, marked by stray edges and appearance of large flattened cells (a). 100× magnified phase contrast image of feeder-dependent PSC with large areas of differentiating cells (b).

cells at the edges of each colony, or in the center, can be mechanically removed with a 25 or 27 gauge needle before passaging (alternatively a P200 tip can be used to scrape away whole differentiated colonies).

It is critical to observe the distribution of attached colonies 24 hours after passage. Colonies sometimes can all be clustered towards the middle, close to each other, or on top of each other. An even distribution of colonies is critical to maintain cells undifferentiated during their expansion. It is also important to monitor the attachment and morphology of the colonies. In all dishes, there will always be some areas with unattached cells or cells that look differentiated. However, the majority of the colonies in the dish must be well separated with a

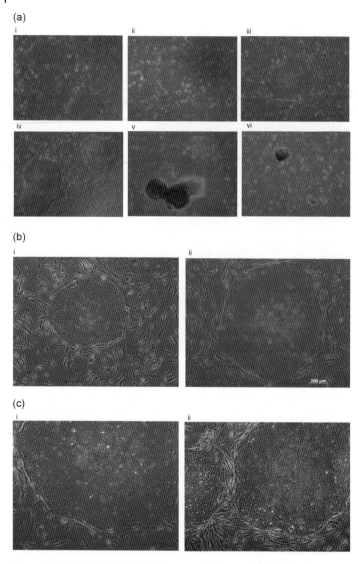

Figure 1.3 Monitoring PSC after passage. 50× magnified phase contrast images of feeder-dependent PSC culture at 24 hours post passage (a), showing colonies with good morphologies (top panels i–iii) and bad morphologies (bottom panels iv–vi). 50× magnified phase contrast images of PSC colonies (b) at day 2 (i) and day 3 (ii) after passaging, showing clear increases in colony size while maintaining morphology. 50× magnified phase contrast images of PSC colonies at day 4 post passage (c) illustrating presence of large colonies (i) and areas with neighboring colonies beginning to grow into each other (ii).

(a) (b)

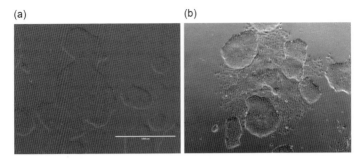

Figure 1.4 Observing feeder-free PSC cultures. 50× magnified phase contrast image of feeder-free PSCs with ideal PSC morphology characterized by sharp edges and compact cells growing as a monolayer (a). 50× magnified phase contrast image of feeder-free PSC with large areas of differentiated cells characterized by heterogeneous central core not typical of pluripotent stem cells and presence of large single cells or small flattened colonies (b).

pristine morphology. Feeder-dependent PSC culture at 24 hours post passage should predominantly show colonies with good morphologies (Figure 1.3a, top panel) and minimal colonies with bad morphology (see Figure 1.3a, bottom panel). At 48–72 hours post passage, well-separated, seeded-down colonies will continue to expand and grow in size while maintaining their morphology (see Figure 1.3b). Feeder-dependent hPSC colonies are typically ready for passage at day 4 or 96 h post passage. At this time, most colonies are large and the neighboring colonies may begin to grow into each other (see Figure 1.3c).

Feeder-free PSCs grow in a monolayer formation without restricted growth. PSCs retain a high nucleus-to-cytoplasm ratio and have tight cell-to-cell junctions. Colony edges are not as defined as in feeder-dependent cultures and colonies may merge as they reach confluency (Figure 1.4a). Differentiating colonies show large differentiated cells and a heterogeneous central core that is not typical of pluripotent stem cells. Single cells and small colonies tend to differentiate, and hence, care is necessary to retain good cluster size for maintaining pluripotency (see Figure 1.4b).

References

1 S. Coecke *et al.* Guidance on good cell culture practice. A report of the second ECVAM task force on good cell culture practice. *Altern Lab Anim* **33**, 261–287 (2005).

2 M.S. Inamdar, L. Healy, A. Sinha, G. Stacey. Global solutions to the challenges of setting up and managing a stem cell laboratory. *Stem Cell Rev* **8**, 830–843 (2012).

2

Pluripotent Stem Cell Culture

2.1 Introduction

The most common types of pluripotent stem cells used are induced pluripotent stem cells (iPSC) or embryonic stem cells (ESC). All PSC are cultured and propagated while striving to maintain the cells in an undifferentiated state. Traditional culture of PSC relies on a supporting layer of feeder cells such as mitotically inactivated mouse embryonic fibroblasts (MEF or iMEF) or human cells such as human foreskin fibroblasts (hFF) [1]. The main concern with PSC culture on mouse feeders is heightened risk of contamination from the animal-derived material, mycoplasma, and potential expression of an immunogenic sialic acid (Neu5Gc) [2]. For human feeders, this risk further includes contamination by viral and non-viral infectious agents [3].

Recent advances have led to successful culture and expansion of PSC under feeder-free conditions using a variety of commercially available media options. Several studies have indicated that genetic and epigenetic stability of PSC are dependent on culture condition and time in culture [4–8]. It is important to understand and implement good cell culture practices while working with stem cells [9].

Human Pluripotent Stem Cells: A Practical Guide, First Edition. Uma Lakshmipathy, Chad C. MacArthur, Mahalakshmi Sridharan and Rene H. Quintanilla.
© 2018 John Wiley & Sons, Inc. Published 2018 by John Wiley & Sons, Inc.

2.2 Materials

All materials are from Thermo Fisher Scientific unless specified otherwise.

2.2.1 Feeder-Dependent PSC Culture

1) DMEM (1X), Liquid (High Glucose) with GlutaMAX™-I *Cat# 10569-010*
2) Mouse (ICR) Inactivated Embryonic Fibroblasts *Cat# A24903*
3) DMEM/F-12 (1X), Liquid (1:1), with GlutaMAX™-I *Cat# 10565-018*
4) Fetal Bovine Serum, ES Cell-Qualified *Cat # 16141-061*
5) MEM Non-Essential Amino Acids Solution (100X) *Cat# 11140-050*
6) Attachment Factor Protein (1X) *Cat# S-006-100*
7) Knockout™ Serum Replacement, KSR *Cat# 10828-010*
8) 2-Mercaptoethanol (1000×), Liquid *Cat# 21985-023*
9) FGF-basic (AA 1-155) Recombinant Human *Cat# PHG0264*
10) Collagenase Type IV *Cat# 17104-019*
11) Dulbecco's Phosphate Buffered Saline (D-PBS) without calcium and magnesium *Cat# 14190-144*
12) Dispase II, powder *Cat# 17105-041*
13) Dimethyl sulfoxide, DMSO *Quality Biological Cat# N1825*
14) StemPro™ EZPassage™ Disposable Stem Cell Passaging Tool *Cat# 23181-010*
15) BD Falcon Cell Scraper *Fisher Cat# 08-771-1A*
16) BD PrecisionGlide Single Use Needle, 27 Gauge *Fisher Cat# 14-821-13B*
17) BD Disposable Syringes *Fisher Cat# 14-823-40*

2.2.2 Feeder-free PSC Culture Using Essential 8 Medium

1) Essential 8™ Medium *Cat# A1517001*
2) Dulbecco's Phosphate Buffered Saline (D-PBS) without calcium and magnesium *Cat# 14190-144*
3) DMEM/F-12 (1X), Liquid (1:1), with GlutaMAX™-I *Cat#10565-018*

4) Geltrex™ LDEV-Free hESC-qualified Reduced Growth Factor Basement Membrane Matrix *Cat# A1413302*
5) Vitronectin (VTN-N) Recombinant Human Protein, Truncated *Cat# A14700*
6) UltraPure™ 0.5M EDTA, pH 8.0, *Cat# 15575020*
7) Versene Solution, *Cat# 15040066*
8) Dulbecco's Phosphate Buffered Saline (D-PBS) **WITH** calcium and magnesium *Cat# 14040133*
9) rh-Laminin 521-Recombinant Human Laminin 521 *Cat# A29248*
10) PSC Cryopreservation Kit *Cat# A2644601*

2.3 Solutions

2.3.1 iMEF Medium (for 500 mL)

DMEM	444.5
FBS (ES qualified)	50 mL
NEAA	5 mL
2-Mercaptoethanol	500 µL

Filter sterilize through a 0.22 µm filter unit. Medium lasts for up to 1 month at 4 °C, protected from light.

2.3.2 Human PSC Medium (for 500 mL)

DMEM-F12	395 mL
KnockOut Serum Replacement	100 mL
NEAA	5 mL
2-Mercaptoethanol	500 µL

Sterilize through a 0.22 µm filter. Medium lasts for up to 28 days at 4 °C. Add bFGF (final concentration 4 ng/mL) fresh daily prior to use (example 0.4 µL reconstituted bFGF per mL of medium).

2.3.3 Basic FGF Solution (10 µg/mL, for 1 mL)

Basic FGF	10 µg
D-PBS (-/-)	990 µL
KnockOut Serum Replacement	10 µL

Aliquot and store at −20 °C for up to 3 months. Once bFGF aliquot is thawed, use within 7 days, when stored at 4 °C.

2.3.4 Collagenase IV Solution (1 mg/mL, for 50 mL)

Collagenase IV	50 mg
DMEM-F12	50 mL

Sterilize through a 0.22 µm filter and store at 4 °C for up to 14 days.

2.3.5 Dispase Solution (2 mg/mL, for 50 mL)

Dispase	100 mg
DMEM-F12	50 mL

Sterilize through a 0.22 µm filter and store at 4 °C for up to 14 days.

2.3.6 hPSC Cryo-Preservation Medium A

Complete hPSC medium (50%) + KSR (50%).

2.3.7 hPSC Cryo-Preservation Medium B

Complete hPSC medium (80%) + DMSO (20%). Make fresh cryo-preservation medium A and B, and store at 4 °C for up to 3 days.

2.3.8 Essential 8 Medium (for 500 mL)

Essential 8 Basal Medium	490 mL
Essential 8 Supplement	10 mL

Sterilize through a 0.22 μm filter. Thaw the frozen Essential 8 Supplement at 2–8 °C overnight or room temperature for 1 hour. Do not thaw the frozen supplement at 37 °C. Store complete Essential 8 medium in a polystyrene bottle at 2–8 °C for up to 14 days, protected from light. Before use, warm the complete medium required for that day at room temperature until it is no longer cool to the touch. Do not warm the medium at 37 °C.

2.3.9 EDTA Passaging Solution (0.5 mM, for 50 mL)

EDTA Ultrapure (0.5 M)	50 μL
D-PBS (-/-)	50 mL

Filter sterilize through a 0.22 μm filter unit. Store at room temperature for up to 4 weeks. It is essential that the EDTA solution be prepared with D-PBS that does not contain calcium chloride or magnesium chloride, as these divalent cations will prevent the EDTA from harvesting the PSC cultures.

2.3.10 Essential 8 Cryopreservation Medium (for 1 mL)

Complete Essential 8 Medium	0.9 mL
DMSO	0.1 mL

Make fresh cryopreservation medium, and store at 4 °C for up to 3 days.

2.4 Methods

2.4.1 iMEF-Coated Vessels for Feeder-Dependent PSC Culture

Feeder-dependent PSC culture requires the preparation of mitotically inactivated (irradiated or mitomycin-c treated) murine iMEFs as a feeder layer, one day before seeding the PSCs onto them during routine passaging. iMEFs must be seeded onto 0.1% gelatin-coated plates (attachment factor is a ready-to-use solution) and must be allowed to attach overnight. iMEF feeder layers secrete extracellular proteins onto the gelatin that permit PSC attachment. In addition, iMEFs secrete growth factors and cytokines that are essential for maintaining PSC pluripotency and growth.

2.4.1.1 Coating Culture Vessels with Attachment Factor

1) Add attachment factor solution (4 °C) into each tissue culture grade dish or each well of a culture plate as per Table 2.1. Ensure that the solution is evenly distributed to ensure even coating.
2) Incubate the dishes or plates for a minimum of 1 hour in a 37 °C, 5% CO_2 incubator to allow for coating before plating iMEFs.

2.4.1.2 Preparing iMEF-Coated Culture Vessels

1) When attachment factor-coated culture vessels are ready, warm appropriate amount of iMEF medium in a 37 °C water bath.
2) Wear eye protection as the cryogenically (−196 °C) stored vials, although optimally stored in the gaseous phase of the liquid nitrogen storage unit, may contain some liquid nitrogen and may accidentally explode when rapidly warmed.
3) Wear ultra-low temperature cryogenic gloves during retrieval of the vials to prevent freezer burn. Remove the necessary cryovials of iMEFs from the liquid nitrogen storage tank using metal forceps.
4) Gently roll the vials (never thaw more than two vials at a time) between your gloved hands for about 5–10 seconds to excess remove frost from the exterior of the vials.

Table 2.1 Volume of attachment factor for iMEF culture.

Vessel	12-well plate	6-well plate	35 mm dish	60 mm dish	100 mm dish	T25 flask	T75 flask
Surface area	$4\,cm^2$	$10\,cm^2$	$10\,cm^2$	$20\,cm^2$	$60\,cm^2$	$25\,cm^2$	$75\,cm^2$
Volume	0.6 mL	1.5 mL	1.5 mL	3 mL	9 mL	3 mL	9 mL

5) Immerse the vials in a 37 °C water bath. Swirl gently to ensure even thawing. *Note*: to prevent contamination, do not submerge the cap into the water.

6) When only an ice crystal remains (usually 2–3 minutes maximum), remove the vials from the water bath.

7) Spray the outside of the vials with 70% ethanol and quickly place in the BSL-2 hood.

8) Quickly transfer the thawed cell suspension into a 50 mL conical tube with a 1 mL pipette (or P1000).

9) Using a 10 mL pipette, add 10 mL of prewarmed iMEF medium to the 50 mL conical tube drop-wise (to avoid osmotic shock to cells), while gently swirling the conical tube to mix the solution evenly. Once all 10 mL of pre-warmed iMEF medium has been added drop-wise, gently pipette up and down to mix the suspension to ensure the cryopreservation media have been thoroughly diluted.

10) Transfer the contents to a 15 mL conical tube and centrifuge at 200× g for 4 minutes.

11) Gently aspirate away the supernatant.

12) Reconstitute the cell pellet and thoroughly mix to obtain a single cell suspension in 1–2 mL iMEF medium.

13) Take a small amount of the cell suspension and count the cell number (cell concentration) and viability using a trypan blue solution and a hemacytometer or the Countess automated cell counter.

14) Aspirate the attachment factor from the culture dishes immediately prior to use (do not allow plates to dry). Add prewarmed iMEF medium into each culture vessel according to Table 2.2.

Table 2.2 Media volume recommendations for seeding of iMEFs.

Vessel	12-well plate	6-well plate	35 mm dish	60 mm dish	100 mm dish	T25 flask	T75 flask
Surface area	$4\,cm^2$	$10\,cm^2$	$10\,cm^2$	$20\,cm^2$	$60\,cm^2$	$25\,cm^2$	$75\,cm^2$
Volume	1 mL	2 mL	2 mL	5 mL	10 mL	5 mL	12 mL

15) Resuspend the iMEF suspension in additional iMEF medium as necessary to add the appropriate amount of cells into each culture vessels according to Table 2.3 in order to achieve a density of 3.0×10^4 cells/cm^2

16) Move the culture vessels in several quick back-and-forth and side-to-side motions to disperse cells across the surface of the culture vessels. Return the culture vessels to the incubator after plating the iMEFs. *Note*: iMEF culture vessels should be used on the next day. Do not plate iMEF more than 2 days prior to use.

17) Before adding PSCs, make sure to wash iMEF dishes with D-PBS twice, after aspirating off the iMEF medium. This will wash away any residual FBS contained in the iMEF medium. *Note*: lot-to-lot variability in iMEF viability and seeding density may be observed. The density of plated-down iMEFs should not be too low or high, but instead should form a uniform coating on the dish surface (Figure 2.1).

Table 2.3 iMEF cell concentrations per vessel configuration.

Vessel	12-well plate	6-well plate	35 mm dish	60 mm dish	100 mm dish	T25 flask	T75 flask
Surface area	4 cm^2	10 cm^2	10 cm^2	20 cm^2	60 cm^2	25 cm^2	75 cm^2
Cell Number	1.2 × 10^5	3 × 10^5	3 × 10^5	6 × 10^5	1.8 × 10^6	7.2 × 10^5	2.2 × 10^6

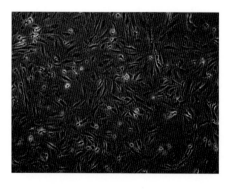

Figure 2.1 50× magnified phase contrast image of iMEFs showing expected density and uniformity of coating for feeder-dependent PSC culture.

2.4.2 Extracellular Matrix Coating
for Feeder-Free PSC Culture

Feeder-independent PSC culture requires extracellular matrix coating of tissue culture-treated vessels for proper cell adherence, growth, and maintenance of pluripotency. Since this type of culture is independent of feeder cells, the extracellular matrix must be coated just prior to usage. Feeder-independent culture is less forgiving and care must be taken to ensure proper growth and attachment. Geltrex LDEV-Free hESC-qualified reduced growth factor basement membrane matrix is a soluble form of basement membrane extracted from murine Engelbreth-Holm-Swarm (EHS) tumors and is free of viruses, including lactose dehydrogenase elevating virus (LDEV). It is a feeder-free matrix that is qualified to support robust PSC cultures (similar to Matrigel) but is not xeno free. Alternative xeno-free, feeder-free matrices include recombinant human vitronectin (VTN) and recombinant human laminin 521, which support feeder-free PSC adherence, cell growth, and pluripotency.

2.4.2.1 Coating PSC Culture Dishes with Geltrex

1) Thaw concentrated Geltrex at 4 °C overnight (protected from light).
2) Remove sterile DMEM/F-12 medium from 4 °C storage. Work aseptically on ice to prevent polymerization of the Geltrex solution. Dilute the concentrated stock Geltrex with an equal amount of DMEM/F-12 (1:1 volume per volume) and mix gently. Aliquot into prechilled (4 °C) sterile microcentrifuge tubes (generally in 0.5 or 1.0 mL aliquots) and freeze immediately at −20 °C. Frozen aliquots are good for up to 3 months. Prior to use, aliquots must be thawed at 4 °C.
3) Using a thawed (4 °C) aliquot, prepare a working stock solution, on ice, diluting the aliquot with additional DMEM/F-12. For general growth, a final dilution of 1:100 is recommended (i.e., dilute a 0.5 mL aliquot with 24.5 mL of DMEM/F-12 medium). Evenly distribute the diluted Geltrex solution to coat the whole surface of each culture plate or dish with the 1:100 Geltrex solution using volumes described in Table 2.4. Thaw enough Geltrex for the scale of each application, and do not refreeze the aliquot.

Table 2.4 Volume of Geltrex for coating of tissue culture vessels.

Vessel	12-well plate	6-well plate	35 mm dish	60 mm dish	100 mm dish	T25 flask	T75 flask
Surface area	$4\,cm^2$	$10\,cm^2$	$10\,cm^2$	$20\,cm^2$	$60\,cm^2$	$25\,cm^2$	$75\,cm^2$
Volume	1 mL	2 mL	2 mL	5 mL	10 mL	5 mL	12 mL

4) Incubate the coated culture vessels for 1 hour at 37 °C in a humidified 5% CO_2 incubator to allow for proper solidification and attachment of the matrix.
5) Geltrex-coated dishes can be used immediately, or up to 1 day after coating. Aspirate off the residual diluted Geltrex solution prior to use and discard. It is not necessary to rinse dishes prior to use. Cells can be passaged directly onto the Geltrex-coated culture dishes.

2.4.2.2 Coating PSC Culture Dishes with Human Recombinant Vitronectin (VTN)

1) Remove a 1 mL vial of vitronectin from −80 °C storage and thaw at room temperature (approximately 30 minutes).
2) Prepare working aliquots by dispensing 60 µL of vitronectin into sterile polypropylene microcentrifuge tubes. The aliquots can be frozen at −80 °C or used immediately.
3) The recommended concentration of vitronectin is $0.5\,\mu g/cm^2$ for the culture vessels, which is a 1:100 (volume/volume) dilution in D-PBS (without $CaCl_2$ and $MgCl_2$) at the time of coating, as per Table 2.5. *Note*: the optimal working concentration of vitronectin is cell line dependent. We recommend using a final coating concentration of $0.1–1.0\,\mu g/cm^2$ on the culture surface, depending on your cell line. We routinely use vitronectin at $0.5\,\mu g/cm^2$ for human PSC culture.
4) Remove a 60 µL aliquot of vitronectin from −80 °C storage and thaw at room temperature for 5 minutes. One 60 µL aliquot is enough to coat one six-well plate or three 60 mm dishes. Thaw enough aliquots for the scale of each application, and do not refreeze.

Table 2.5 Vitronectin volume recommendations for PSC culture.

Vessel	12-well plate	6-well plate	35 mm dish	60 mm dish	100 mm dish	T25 flask	T75 flask
Surface area	$4\,cm^2$	$10\,cm^2$	$10\,cm^2$	$20\,cm^2$	$60\,cm^2$	$25\,cm^2$	$75\,cm^2$
Volume	0.4 mL	1 mL	1 mL	2 mL	6 mL	2.5 mL	7.2 mL

5) Add 60 µL of thawed vitronectin into a 15 mL conical tube containing 6 mL of sterile D-PBS without $CaCl_2$ and $MgCl_2$ at room temperature. Gently mix the diluted vitronectin by pipetting up and down. *Note*: this results in a working concentration of 5 µg/mL (i.e., a 1:100 dilution). Aliquot 1 mL of the diluted vitronectin solution to each well of a six-well plate (scale to TC dish or plate as indicated in Table 2.5).

6) Incubate the coated dishes/plates in a 37 °C, 5% CO_2 humidified incubator for 1 hour. *Note*: do not allow the vessels to dry.

7) Just prior to use, aspirate off the residual diluted vitronectin solution from the culture vessel and discard. It is not necessary to rinse off the culture vessel after removal of vitronectin. Cells can be passaged directly onto the VTN-coated culture dishes.

2.4.2.3 Coating PSC Culture Dishes with Recombinant Human Laminin 521

1) Remove a 1 mL vial of rh-laminin 521 from −20 °C storage and thaw at 4 °C until fully thawed (overnight). Thawed rh-Laminin-521 can be kept at 4 °C for up to 3 months.

2) When fully thawed, dilute the stock 100 µg/mL stock rh-laminin 521 to a working concentration that is optimized for your cell line and application. We recommend a coating concentration of 0.5 µg/cm².

3) Dilute the concentrated stock rh-Laminin 521 solution in the appropriate amount of D-PBS (+$CaCl_2$ and $MgCl_2$) to a final working concentration of 2.5 µg/mL, as per Table 2.6.

4) Add the diluted laminin 521 to each well of the TC plate or TC dish and insure the surface is thoroughly coated. Incubate the coated culture vessels for 2 hours at 37 °C in a humidified

Table 2.6 Laminin 521 volume recommendations for PSC culture (2.5 µg/mL).

Vessel	12-well plate	6-well plate	35 mm dish	60 mm dish	100 mm dish	T25 flask	T75 flask
Surface area	4 cm^2	10 cm^2	10 cm^2	20 cm^2	60 cm^2	25 cm^2	75 cm^2
Volume	0.4 mL	2 mL	2 mL	4 mL	6 mL	5 mL	15 mL

CO_2 incubator or overnight at 4 °C. Once coated, use the vessels right away and do not allow them to dry out.

5) Prior to use, aspirate off the residual rh-laminin 521 solution and discard. It is not necessary to rinse off the culture vessel after the removal of rh-laminin 521. Cells can be passaged directly onto the rh-laminin 521-coated culture vessels.

2.4.3 Thawing and Plating Human PSC on iMEF

1) The day prior to thawing hPSC, make the appropriate iMEF plates or dishes. Typically, we recommend recovering a vial of cells onto a single 60 mm dish of iMEF. Scale up or down based on cell density of PSCs at the time of freezing.

2) Thirty minutes prior to thaw, warm 15 mL of human PSC medium (KnockOut Serum Replacement-containing medium) in a 37 °C water bath.

3) Prior to thaw, aspirate off the iMEF medium from a 60 mm dish containing inactivated MEFs and wash the dish with 5 mL of room-temperature D-PBS twice to remove any residual FBS from the iMEF-coated dish.

4) Aspirate off the D-PBS wash and add 3 mL of prewarmed hPSC medium (supplemented with fresh bFGF to a final concentration of 4 ng/mL) to the dish prior to plating the thawed PSCs. Keep the dish in a 37 °C, humidified CO_2 incubator until use.

5) Wear ultra-low temperature cryogenic gloves during the retrieval of the vials to prevent freezer burn. Remove the necessary cryovials of iMEFs from the liquid nitrogen storage tank using metal forceps.

6) Gently roll the vials (never thaw more than two vials at a time) between your gloved hands for about 5–10 seconds to excess remove frost from the exterior of the vials.

7) Immerse the vials in a 37 °C water bath. Swirl gently to ensure even thawing. Note: to prevent contamination, do not submerge the cap into the water.

8) When only an ice crystal remains (usually 2–3 minutes maximum), remove the vials from the water bath.

9) Spray the outside of the vials with 70% ethanol and quickly place in the BSL-2 hood.

10) Transfer cells gently into a sterile 50 mL conical tube using a 5 mL pipette.

11) Using a 10 mL pipette, add 10 mL of prewarmed hPSC medium to the 50 mL conical tube drop-wise. While adding the medium, gently move the tube back and forth to mix the PSCs. This reduces osmotic shock to the PSCs.

12) Transfer the contents to a 15 mL conical tube and centrifuge the cells at 200× g for 2 minutes.

13) Gently aspirate off the supernatant.

14) Resuspend the cell pellet in 2 mL of hPSC medium using a 5 mL pipette by gently pipetting up and down (2–3 times, being careful to not further break up the cell clusters).

15) Label the iMEF culture dish with the nomenclature of the PSC cell line, the appropriate passage number (typically one additional passage number is added to that from freeze; for example, if frozen at P22, cells are labeled P23 at thaw) based on the information on the vial and the date of thaw, along with the user's initials.

16) Slowly add the PSC suspension drop-wise, while gently swirling, into the iMEF dish previously prepared.

17) Move the dish gently in several figure-of-eight motions to disperse the PSC colonies across the surface of the dish. Place the dish gently into the incubator, and repeat the dispersal. *Note*: PSC clusters seeded onto iMEF-coated dishes must show uniform-sized colonies that are evenly distributed to prevent colonies from clumping together and attaching too close to each other (Figure 2.2).

18) The following day after thaw, gently aspirate off the medium and any unattached cell clusters. Replace the spent medium with fresh prewarmed hPSC medium, freshly supplemented with bFGF. If feeding more than one dish, use a different pipette for each dish to reduce risk of contamination.

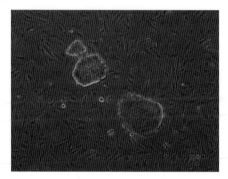

Figure 2.2 50 × magnified phase contrast image of uniform-sized PSC clusters right after thaw, seeded onto iMEF-coated dishes.

19) Replace the medium daily (feed) for no more than 7 continuous days prior to splitting the cells from thaw.
20) Observe PSCs daily and passage the cultures whenever the colonies are too big or crowded. The ratio of splitting depends on the total number of PSC colonies in the culture vessels (approximately 1:1 to 1:3 for PSCs at the first passage after recovery, depending on recovery confluency, and then adjust thereafter).

2.4.4 Thawing and Plating Human PSC on Essential 8 Medium

1) Place 10 mL of Essential 8 Medium in a 15 mL tube and warm to room temperature. *Note*: do not warm medium in a 37 °C water bath.
2) Generally thaw one vial of hPSC in a 60 mm dish (scale up or down as necessary as per the density of the PSC at the time of freezing).
3) Wear ultra-low temperature cryogenic gloves during the retrieval of the vials to prevent freezer burn. Remove the cry-ovial of PSCs from the liquid nitrogen storage tank using metal forceps.
4) Gently roll the vial between your gloved hands for about 5–10 seconds to excess remove frost from the exterior of the vial.
5) Immerse the vials in a 37 °C water bath. Swirl gently to ensure even thawing. *Note*: to prevent contamination, do not submerge the cap into the water.

6) When only an ice crystal remains (usually 2–3 minutes maximum), remove the vials from the water bath.

7) Spray the outside of the vials with 70% ethanol and quickly place in the BSL-2 hood.

8) Transfer cells gently into a sterile 50 mL conical tube using a 5 mL pipette.

9) Slowly add 10 mL of Essential 8 medium drop-wise to the cells in the 50 mL conical tube. While adding the medium, gently move the tube back and forth to mix the PSCs. This reduces osmotic shock to the cells.

10) Transfer the contents to a 15 mL conical tube and centrifuge the cells at 200× g for 2 minutes.

11) Aspirate and discard the supernatant.

12) Resuspend the cell pellet in 2 mL of Essential 8 medium by gently pipetting the cells up and down in the tube 2–3 times (be very gentle so as not to further break up the cell clusters).

13) Aspirate off the residual diluted Geltrex or vitronectin from the coated dish. Add 3 mL of Essential 8 medium to the coated dish.

14) Slowly add, drop-wise, the 2 mL PSC suspension into the coated dish.

15) Move the dish in several quick figure-of-eight or side to side, back and forth motions, to disperse cells across the surface of the wells.

16) Place dish gently into the 37 °C, 5% CO_2 incubator and repeat the dispersion of the cells to ensure even distribution.

17) Incubate the cells overnight in the incubator.

18) The next day, aspirate off the spent medium and replace (feed) with fresh complete Essential 8 medium.

19) Replace the medium daily thereafter until the cells are approximately 85% confluent prior to passaging them onto freshly coated dishes.

2.4.5 Boosting Recovery of Cryopreserved PSC

Often PSCs, especially early passage iPSCs and difficult to recover iPSCs, require assistance in recovery from cryopreservation. RevitaCell Supplement is a chemically defined recovery supplement containing a specific ROCK inhibitor as well as molecules that have antioxidant and free radical scavenger properties. Use of RevitaCell Supplement during thaw can help with recovery of difficult PSC lines.

1) Follow the appropriate thawing procedure for your PSCs of choice, either feeder dependent or feeder free as previously described.
2) Centrifuge the cell suspension at 200× g for 2 minutes.
3) Aspirate the medium, being careful not to disturb the cell pellet.
4) Gently resuspend the cells in growth medium supplemented with RevitaCell Supplement at a 1× final concentration (i.e., 100 μL of RevitaCell Supplement in 10 mL of growth medium). *Note*: do not add any additional ROCK inhibitors to the growth medium.
5) Transfer the cell suspension to the culture vessel precoated with the appropriate substrate. Move the vessel in several quick back and forth and side to side motions to disperse the cells across its surface.
6) Incubate the cells for 18–24 hours in the recommended cell culture environment.
7) Following incubation, aspirate the growth medium supplemented with RevitaCell Supplement and replace it with unsupplemented growth medium (i.e., without the addition of RevitaCell Supplement) for the remainder of the culture.

2.4.6   Passaging hPSC

In general, split cells when one of the following occurs.

- iMEF or feeder-free matrix (e.g., Geltrex or VTN) is 7 days old.
- PSC colonies are becoming too dense or too large.
- PSC colonies are showing increased differentiation.
- The colonies cover approximately 85% of the surface area of the culture vessel.

Pluripotent stem cells are usually split every 4 days. The split ratio can vary, though it is generally between 1:2 and 1:4 for early passages lines and coming out of thaw. For established cultures, the split ratio is typically between 1:3 and 1:6 for feeder-dependent culture, and between 1:6 and 1:12 for feeder-free cultures. Occasionally, cells will grow at a different rate and the split ratio will need to be adjusted. A general rule is to observe the last split ratio and adjust the ratio according to the appearance of the PSC colonies. If the cells look healthy and colonies have enough space, split using the same ratio. If they are overly dense and crowding, increase the ratio. If the cells are sparse, decrease the ratio.

2.4.6.1   Enzymatic Passage of Human PSCs

Enzymatic passaging of PSCs will vary from cell line to cell line. Some PSC cell lines may not react in the same manner to enzymatic passaging. Consequently, the type of enzyme utilized, the concentration of enzyme, and exposure time must be empirically determined for the particular cell line to be passaged. Here we describe the typical enzymatic passaging procedure using collagenase IV, for hESC. Note that dispase can also be used and is noted where appropriate. If the hESC or iPSC line being cultured is not optimally passaged enzymatically, manual or mechanical passaging must be performed. When PSC colonies are harvested in bulk using enzymatic methods such as treatment with collagenase IV (Figure 2.3), attached PSC colonies curl up, then detach from the dish, leaving behind iMEFs. When the colonies begin to curl up, cell clusters can be gently dislodged with a 5 mL pipette.

1) Prepare new iMEF plates or dishes, one or two days before PSC splitting.
2) Aspirate off the spent hPSC medium from the confluent PSC culture dish. Add 2 mL of 1× collagenase IV solution to a 60 mm dish containing PSCs. Adjust the volume of collagenase IV for various dish sizes, as per Table 2.7.
3) Incubate the PSC cultures 30–60 minutes in a 37 °C, 5% CO_2 incubator. *Note*: incubation times may vary among different batches of collagenase and different types of PSC, so you need to determine the appropriate incubation time by examining the colonies.
4) *Alternative*: you may use dispase at a concentration of 2 mg/mL instead of collagenase IV. Incubate 5–15 minutes in a 37 °C, 5% CO_2 incubator.

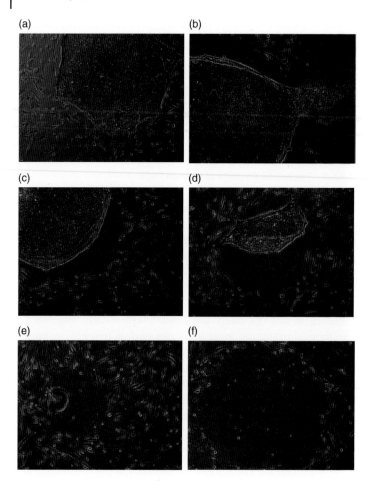

Figure 2.3 50× magnified phase contrast images of PSC colonies harvested in bulk using collagenase IV for 30–60 minutes, showing different stages of PSC colony detachment. First the edges of the colony will appear to thicken and start moving inwards (a). As incubation time increases, the edges will continue to move inwards (b,c), and then eventually the colony will curl up and detach from the dish (d,e), leaving behind iMEFs (f). The ideal time to harvest colonies is when the colonies just begin to curl up (b,c), and they should be gently dislodged with a 5 mL pipette.

Table 2.7 Volume of collagenase IV for PSC enzymatic passaging.

Vessel	12-well plate	6-well plate	35 mm dish	60 mm dish	100 mm dish	T25 flask	T75 flask
Surface area	$4\,cm^2$	$10\,cm^2$	$10\,cm^2$	$20\,cm^2$	$60\,cm^2$	$25\,cm^2$	$75\,cm^2$
Volume	0.4 mL	1 mL	1 mL	2 mL	5 mL	2 mL	6 mL

5) Stop the incubation when the edges of the colonies start to pull away from the plate.
6) After incubation, gently dislodge the colonies with a 5 mL pipette by gently pipetting colonies across the surface of the plate 5–8 times. This breaks up the colonies into smaller clumps.
7) Transfer the 2 mL collagenase solution/colony suspension to a 15 mL conical tube. Add 2 mL of prewarmed hPSC medium to the original dish and pipette across the surface of the dish to dislodge any remaining colonies. Transfer this 2 mL suspension to the 15 mL conical tube and pipette up and down another 2–3 times to resuspend colonies. Be sure not to introduce bubbles or shear colonies too much. You want to break up the colonies into small clusters (50–500 cells) for normal passaging.
8) Centrifuge the cells at 200× g for 2 minutes at room temperature. Alternatively, you can let the colonies settle to the bottom of the tube via gravity by allowing the tube to stand at room temperature for 5–10 minutes. Gravity sedimentation will allow desired PSC colonies to settle while any iMEFs, differentiated cells, dead cells, single cells, and undesirably small colony fragments will not settle down and can be aspirated.
9) Aspirate and discard the supernatant, and then gently tap the tube to loosen the cell pellet from the bottom of the tube.
10) Add 2–5 mL of prewarmed hESC medium and resuspend the colonies by gently pipetting up and down. *Note*: at this point, avoid breaking up the cell clusters any further.
11) Seed the clusters of cells onto a dish plated with iMEFs. Generally PSCs are passaged at a 1:4 to 1:6 split ratio. The final volume of medium depends on the plates used (Table 2.8).

Table 2.8 Media volume recommendations for PSC culture.

Vessel	12-well plate	6-well plate	35 mm dish	60 mm dish	100 mm dish	T25 flask	T75 flask
Surface area	4 cm^2	10 cm^2	10 cm^2	20 cm^2	60 cm^2	25 cm^2	75 cm^2
Volume	1 mL	2 mL	2 mL	5 mL	10 mL	5 mL	15 mL

12) Place the plates in a 37 °C, 5% CO_2 incubator. Shake the plates gently, by swirling them in several figure-of-eight motions, to evenly spread out the cell clusters.

2.4.6.2 Mechanical Passaging of PSC

In the overall PSC workflow, sometimes it may be necessary to manually dissect PSC colonies to either remove undesired portions or whole colonies, or to break up and passage individual colonies. Traditional dissection tools include pulled glass Pasteur pipettes and fine gauge hypodermic needles. In particular, this method is used for maintenance of PSC colonies by removing unwanted differentiating colonies, and for manually passaging colonies of PSCs that cannot be passaged enzymatically. Mechanical passaging is also an important tool for hiPSC selection and maintenance. Here we describe protocols for traditional mechanical passaging of individual colonies and for bulk passaging of PSC culture using the StemPro EZPassage Disposable Stem Cell Passaging Tool.

2.4.6.2.1 Manual Passaging of PSC Using a Needle and Syringe

Note: manual passaging can only be performed in TC dishes and is not intended for PSCs grown in flasks as the use of a needle cannot readily access individual colonies in a flask.

1) For optimal dissection of individual colonies, you must be able to work aseptically under a laminar flow hood using sterile materials. Consequently, a dissection stereomicroscope within the hood is preferred. As a stereomicroscope will require a modified BSL-2 sash to permit the oculars to be accessed from outside the hood, while manipulations are performed inside the hood, alternatively we recommend

using a microscope with a display that can be kept inside the hood, such as the EVOS imaging systems (XL, XL core, or FL).

2) First focus on the desired colony to dissect. Using a single-use 25 to 27 gauge needle, bevel up with an attached 3 mL syringe, make long parallel straight cuts across the colony. Rotate the TC dish 90° and make the same parallel cuts into the colony. This will make small squared clusters of cells for passaging purposes (Figure 2.4).

3) Using a sterile P200 pipettor, scrape up the individual clusters with the sterile pipette tip and aspirate the floating cluster. Transfer the cell clusters to a fresh culture dish for continued growth. Repeat the procedure for desired colonies.

4) For bulk passaging of PSCs, score across the whole plate and collect cell clumps using a cell scraper. Split as desired or as described previously.

5) For dissecting out undesired portion of the colonies during normal maintenance, use the same procedure using a 25–27 gauge needle (a P200 tip can also be used) to score around the undesired portion and cut out and dispose of the any differentiated cells prior to passage.

(a)　　　　　　　　　　　　(b)

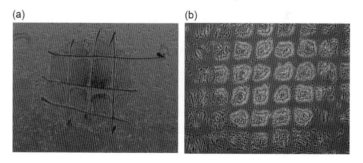

Figure 2.4 50× magnified phase contrast image shows a PSC colony manually dissected using a needle and syringe (a). Image shows straight cuts in horizontal and vertical directions creating a checkerboard pattern of small cell clusters ready to be harvested, and then reseeded for expansion and passaging. 100× magnified phase contrast image of a PSC colony where the entire dish was scored using the StemPro EZPassage tool; the resulting clumps of the colony are ready to be harvested and transferred to a fresh iMEF-coated dish (b).

2.4.6.2.2 Manual Passaging of PSC using StemPro EZPassage Disposable Stem Cell Passaging Tool

Note: this protocol is not for PSCs cultured in flasks because it is difficult to remove differentiated colonies in a flask. Also, the StemPro EZPassage disposable stem cell passaging tool and cell scraper cannot reach colonies in a flask.

1) Prepare the new iMEF plates or dishes one or two days before PSC passaging.
2) Remove the PSC plates or dishes from the incubator. Label differentiated colonies with a microscope marker and remove differentiated colonies with a P200 pipette tip or needle by scraping them off under a microscope in a BSL-2 hood as previously described.
3) Aspirate off the spent PSCs prior to any further manipulations to avoid carry-over of differentiated cells.
4) Add 1 mL prewarmed hESC medium to each well of a six-well plate, 2 mL to each 60 mm dish or 4 mL to each 100 mm dish.
5) Roll the StemPro EZPassage disposable stem cell passaging tool across the entire dish or plate in one direction (left to right). Rotate the culture dish or plate 90°, and roll again across the entire dish or plate. This procedure produces relatively uniform-sized cell clumps (see Figure 2.4).
6) Use a cell scraper to gently detach the cells off the surface of the plates or dishes.
7) Gently transfer cell clumps using a 5 mL pipette and place into a 15 mL conical tube. Note: do not break cell clumps any further.
8) Add 1 mL prewarmed human PSC medium to each well of a six-well plate, 2 mL to each 60 mm dish or 3 mL to each 100 mm dish to collect residual cells and add to the cell suspension in the conical tube.
9) Aspirate the medium from plates or dishes containing iMEF. Wash the plates with D-PBS to remove residual FBS.
10) Add prewarmed PSC medium to each culture dish as described in Table 2.8.
11) Allow the cell clusters in the conical tube to gravity sediment for 5–10 minutes in the BSL-2 hood. (Alternatively, centrifugation at 200× g for 2 minutes can be used to harvest the cell clusters.)

12) Once a cell cluster pellet has formed, gently aspirate off the supernatant without disturbing the pellet.

13) Add the appropriate amount of medium to the cell clusters to obtain the desired split ratio. Gently resuspend the cell clusters by pipetting up and down with a 5 mL pipettor. Add the appropriate amount of PSC suspension into each well of a culture plate or dish according to the desired split ratio.

14) Label the new vessels with the cell line name, the new passage number, the date, the split ratio, and your initials.

15) Return the vessels to the incubator after plating PSCs. Move the culture vessels in several figure-of-eight motions to disperse cells across the surface of the vessels.

16) Incubate the culture vessels overnight to allow colonies to attach. Replace spent medium daily. Observe PSC cultures every day and passage cells whenever the colonies are too big or crowded (approximately every 3–4 days).

2.4.6.3 Feeder-Free Passaging of Human PSCs on Essential 8 Medium

For cultures on Essential 8 medium, newly derived PSC lines may contain a fair amount of differentiation through passage 4. It is not necessary to remove differentiated material prior to passaging. By propagating/splitting the cells, the overall culture health should improve throughout the early passages. Enzymes such as collagenase and dispase do not work well with cells cultured in Essential 8 medium and vitronectin. Use of these enzymes for passaging cells results in compromised viability and attachment. Cells grown in Essential 8 medium on Geltrex, vitronectin, or laminin must be harvested using EDTA to ensure proper clump size for routine passaging. Washes with D-PBS (without calcium and magnesium) are vital to remove residual divalent cations prior to passaging. Cell harvesting can be performed at either room temperature or 37 °C, but exposure times must be monitored and adjusted depending on the matrix used, and to account for cell line variation (Figure 2.5).

2.4.6.3.1 *Passaging PSC Colonies Using EDTA*

1) Prewarm the required volume of Essential 8 medium at room temperature until it is no longer cool to the touch. *Note*: do not warm medium in a 37 °C water bath.

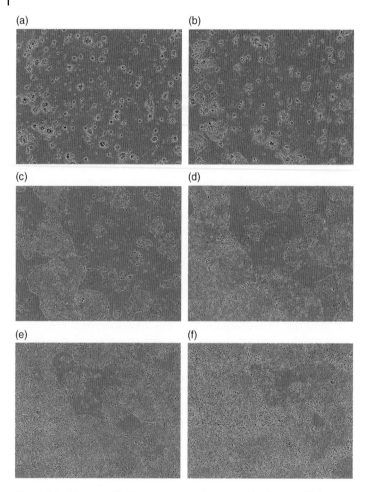

Figure 2.5 50× magnified phase contrast images of feeder-free PSCs grown in Essential 8 medium and vitronectin. Images are of cells immediately after passaging and seeding onto a fresh VTN-coated plate (a), and at days 1–5 (b–f) after seeding.

2) Aspirate off the spent Essential 8 medium from the vessel containing the PSCs.
3) Add the appropriate amount of room-temperature D-PBS without calcium and magnesium to the culture dish. Refer to Table 2.9 for the recommended volumes.

Table 2.9 D-PBS and EDTA volume recommendations for PSC culture.

Vessel	12-well plate	6-well plate	35 mm dish	60 mm dish	100 mm dish	T25 flask	T75 flask
Surface area	$4\,cm^2$	$10\,cm^2$	$10\,cm^2$	$20\,cm^2$	$60\,cm^2$	$25\,cm^2$	$75\,cm^2$
D-PBS volume	1 mL	2 mL	2 mL	5 mL	10 mL	5 mL	15 mL
EDTA volume	0.4 mL	1 mL	1 mL	2 mL	6 mL	2.5 mL	5 mL

4) Aspirate off the first D-PBS wash, and repeat.
5) Aspirate off the second D-PBS wash and add the appropriate amount of 0.5 mM EDTA passaging solution to the vessel containing PSCs. Adjust the volume of EDTA for various dish sizes. Refer to Table 2.9. Swirl the dish to coat the entire cell surface.
6) Incubate the vessel at room temperature for 5–8 minutes or at 37 °C for 3–5 minutes. When the cells start to separate and round up, and the colonies appear to have holes in them when viewed under a microscope, they are ready to be removed from the vessel. *Note*: in larger vessels or with certain cell lines, this may require longer incubation times.
7) Aspirate off the EDTA solution, being careful not to aspirate off the cell clusters.
8) Add prewarmed complete Essential 8 medium to the new, freshly matrix coated dishes according to Table 2.10.
9) Remove the cells from the culture dishes or plates by gently pipetting 2 mL of Essential 8 medium across the surface of

Table 2.10 Essential 8 medium volume recommendations for PSC culture.

Vessel	12-well plate	6-well plate	35 mm dish	60 mm dish	100 mm dish	T25 flask	T75 flask
Surface area	$4\,cm^2$	$10\,cm^2$	$10\,cm^2$	$20\,cm^2$	$60\,cm^2$	$25\,cm^2$	$75\,cm^2$
Volume	1 mL	2 mL	2 mL	5 mL	10 mL	5 mL	15 mL

the plate using a 5 mL pipette. Rotate the plate and repeat to flush off cells from the matrix. Avoid creating bubbles. *Note*: do not overtriturate to prevent the clusters from becoming too small.

10) Collect the cell clusters in a 15 mL conical tube. Add another 2 mL of Essential 8 medium to the dish to harvest any remaining cells. *Note*: do not scrape the cells from the dish. There may be obvious patches of cells that were not dislodged and left behind. Do not attempt to recover them through scraping. Little or no extra pipetting is required to break up cell clumps after EDTA treatment. Work quickly to remove cells after adding Essential 8 medium to the vessels during harvest. The initial effect of the EDTA will be neutralized quickly by the medium. Some lines readhere very rapidly after medium addition, while others are slower to reattach.

11) Add more Essential 8 medium to the cell suspension to obtain the correct volume to perform the desired split ratio. Typically, a 1:6 to 1:12 split ratio is used for routine culture.

12) Add the appropriate amount of cell suspension to the new plates containing medium and move the vessel in several quick figure-of-eight or side to side, back and forth motions to disperse cells across the surface of the vessels.

13) Place the vessel gently into a 37 °C, 5% CO_2 humidified incubator and incubate the cells overnight.

14) After overnight incubation, aspirate off the spent medium and replace with fresh prewarmed Essential 8 medium. *Note*: it is normal to see cell debris and small colonies after passage.

15) Replace the spent Essential 8 medium daily until the cells reach 70–80% confluency, at which time they are ready to be passaged again.

2.4.6.3.2 Adaptation of Feeder-Dependent Cultures to Feeder-Free Conditions Using Essential 8 Medium

Note: the volumes given in the following adaptation procedure are for 60 mm culture dishes. For culture vessels with different sizes, adjust the volumes appropriately.

1) Prepare 1:100 Geltrex matrix solution (for general applications) in DMEM/F-12 medium or 1:50 vitronectin solution (for xeno-free applications) in D-PBS without calcium and magnesium. Coat your culture dishes with your matrix of choice and incubate for 1 hour at 37 °C.

2) Aspirate off the spent PSC medium from the dish containing PSCs on feeder cells and wash once with D-PBS.

3) Aspirate off the D-PBS wash, and add 2 mL of 1× collagenase type IV (1 mg/mL), prewarmed to 37 °C. *Note*: alternatively, you can use dispase II solution (2 mg/mL) in place of the collagenase IV solution.

4) Incubate the cells for 30–60 minutes in a 37 °C, 5% CO_2 incubator. *Note*: if using dispase II solution (2 mg/mL), incubation time is 15–25 minutes in a 37 °C, 5% CO_2 incubator.

5) Stop the incubation when the edges of the colonies begin to curl from the plate.

6) Add 2 mL of complete PSC medium and gently dislodge the colonies from the plate by washing them off using a 5 mL serological pipette. Tip the plate at a 45° angle and rotate the plate as you begin to triturate the clusters of colonies into smaller fragments. Repeat until the desired fragment size is achieved. *Note*: Optimal fragment size for the colonies is critical for successful adaptation. Colony fragments that are too large will form EB-like clusters when reseeded, and fragments that are too small will differentiate.

7) Transfer the suspended colony clusters into a 15 mL conical tube.

8) Add another 2 mL of complete PSC medium to dislodge the remaining colonies and transfer them to the 15 mL tube. *Note*: if desired, you can triturate the PSC colonies by pipetting them up and down in the 15 mL conical tube rather in the tissue culture dish.

9) Let the colony fragments sediment at the bottom of the 15 mL tube for 5–10 minutes by gravity.

10) Discard the supernatant, add 4 mL of complete PSC medium, and gently resuspend the colony fragments by pipetting up and down twice.

11) Gravity sediment the clusters again for 5–10 minutes.

12) While colony fragments are sedimenting, aspirate off the matrix solution from the freshly prepared dishes and add 4 mL of complete Essential 8 medium.
13) Aspirate off the supernatant from the sedimented clusters and add 4 mL complete of Essential 8 medium.
14) Resuspend the PSC clusters by gently pipetting them up and down twice, taking care not to further break down the clusters; the goal is to just resuspend the PSC clusters for seeding (split ratio 1:4).
15) Distribute 1 mL of the suspended PSC clusters into each matrix-coated dish containing 4 mL complete of Essential 8 medium. Gently shake each dish back and forth in both directions for uniform distribution of the clusters.

A double sedimentation is *critical* for attachment of the colonies adapted onto Geltrex or vitronectin-coated plates. Once adapted, cultures can be passaged under normal feeder-independent conditions (see section 2.4.6.1) (Figure 2.6).

2.4.7   Cryopreservation of hPSC

2.4.7.1   Cryopreserving hPSCs Cultured on iMEF

1) Warm the appropriate amount of collagenase IV and hPSC medium to 37 °C in a water bath.
2) Aspirate off the spent culture medium from culture vessels containing confluent PSCs. Add the collagenase IV into each well of the culture vessels according to Table 2.7. Incubate for 30–60 minutes at 37 °C.
3) Stop the incubation when the edges of the colonies start to pull away from the plate.
4) After incubation, gently dislodge the colonies with a 5 mL pipette by gently pipetting colonies up and down, across the surface of the plate 5–8 times. This will also help break up the colonies into smaller clumps.
5) Transfer the 2 mL collagenase solution/colony suspension to a 15 mL conical tube. Add 2 mL of prewarmed hPSC medium to the original dish and pipette across the surface of the dish to dislodge any remaining colonies. Transfer this 2 mL suspension to the 15 mL conical tube and pipette up and down another 2–3 times to resuspend colonies. *Note*: be sure not to

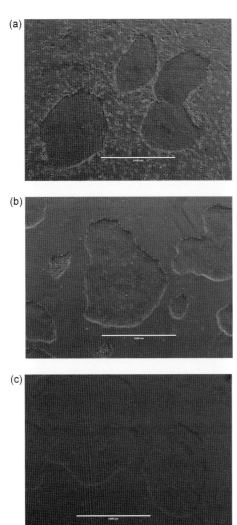

Figure 2.6 50× magnified phase contrast images of PSCs from the same starting population, adapted using the double sedimentation protocol from feeder-based cultures into different feeder-free media and matrices, including KSR-based medium and iMEFs (a), Essential 8 medium and vitronectin (b), and Essential 8 medium and Geltrex (c).

introduce bubbles or shear colonies too much. You want to break up the colonies into small clusters (50–500 cells) for freezing. Alternatively, you can collect cell clusters by using the StemPro EZPassage tool (see section 2.4.6.2.2).

6) Centrifuge the cells at 200× g for 2 minutes at room temperature. Alternatively, you can let the colonies settle to the bottom of the tube via gravity by allowing the tube to stand at room temperature for 5–10 minutes. Gravity sedimentation will allow desired PSC colonies to settle while any iMEFs, differentiated cells, dead cells, single cells, and undesirably small colony fragments will not settle down and can be aspirated.

7) Gently aspirate off the supernatant and resuspend the cells with 0.5 mL (per 60 mm dish of harvested PSCs) of hPSC cryopreservation medium A using a 5 mL pipette.

8) Add the same volume of hPSC cryopreservation medium B, drop-wise, while gently shaking the conical tube between drops. You must work quickly at this time to minimize the cells' exposure to DMSO prior to freezing.

9) Allocate 1 mL cell suspension into each cryogenic vial and place vials in an isopropanol freezing container (e.g., Mr. Frosty) and transfer them to –80 °C overnight.

10) The following day, transfer the cryogenic vials into a liquid nitrogen tank for long-term storage.

2.4.7.2 Cryopreserving Human PSCs Cultured in Essential 8 Medium

1) Prepare Essential 8 cryopreservation medium fresh before use and place the tube on ice until use. Discard any remaining freezing medium after use. *Note*: alternatively, PSC cryopreservation medium (a component of the PSC cryopreservation kit) can be used instead of Essential 8 cryopreservation medium. This medium should be thawed at 4 °C, and is ready to use.

2) Aspirate off the spent Essential 8 medium from the dishes to be harvested, and rinse the cells twice with D-PBS without calcium and magnesium (see Table 2.9).

3) Add 0.5 mM EDTA passaging solution to the dish. Adjust the volume of EDTA for various dish sizes (see Table 2.9). Swirl the dish to coat the entire cell surface.

Table 2.11 Cryopreservation medium volume recommendations for cryopreserving feeder-free PSCs.

Vessel	12-well plate	6-well plate	35 mm dish	60 mm dish	100 mm dish	T25 flask
Surface area	$4\,cm^2$	$10\,cm^2$	$10\,cm^2$	$20\,cm^2$	$60\,cm^2$	$25\,cm^2$
Volume	2 mL	2 mL	3 mL	9 mL	3 mL	9 mL

4) Incubate the dish at room temperature for 5–8 minutes or 37 °C for 3–5 minutes. When the cells start to separate and round up, and the colonies appear to have holes in them when viewed under a microscope, they are ready to be removed from the vessel.

5) Aspirate off the EDTA solution.

6) Add an appropriate amount (Table 2.11) of ice-cold Essential 8 cryopreservation medium (or PSC cryopreservation medium) to the dish.

7) Remove the cells by gently squirting the colonies/freezing media from the plate. Avoid creating bubbles.

8) Collect cells in a 15 mL conical tube on ice. Pool cells from multiple wells if needed.

9) Resuspend cells gently. Aliquot 1 mL of the cell suspension into each cryovial.

10) Quickly place the cryovials in an isopropanol freezing container (e.g., Mr. Frosty) and transfer them to −80 °C overnight.

11) After overnight storage at −80 °C, transfer the cells to a liquid nitrogen tank vapor phase for long-term storage.

References

1 P.H. Schwartz, D.J. Brick, H.E. Nethercott, A.E. Stover. Traditional human embryonic stem cell culture. *Methods Mol Biol* **767**, 107–123 (2011).

2 M.J. Martin, A. Muotri, F. Gage, A. Varki. Human embryonic stem cells express an immunogenic nonhuman sialic acid. *Nat Med* **11**, 228–232 (2005).

3 G.N. Stacey *et al.* The development of 'feeder' cells for the preparation of clinical grade hES cell lines: challenges and solutions. *J Biotechnol* **125**, 583–588 (2006).

4 I. Garitaonandia *et al.* Increased risk of genetic and epigenetic instability in human embryonic stem cells associated with specific culture conditions. *PLoS One* **10**, e0118307 (2015).

5 S.E. Peterson, J.F. Loring. Genomic instability in pluripotent stem cells: implications for clinical applications. *J Biol Chem* **289**, 4578–4584 (2014).

6 Q. Bai *et al.* Temporal analysis of genome alterations induced by single-cell passaging in human embryonic stem cells. *Stem Cells Dev* **24**, 653–662 (2015).

7 A. Maitra *et al.* Genomic alterations in cultured human embryonic stem cells. *Nat Genet* **37**, 1099–1103 (2005).

8 M.M. Mitalipova *et al.* Preserving the genetic integrity of human embryonic stem cells. *Nat Biotechnol* **23**, 19–20 (2005).

9 D. Pamies *et al.* Good Cell Culture Practice for stem cells and stem-cell-derived models. *ALTEX* **34**, 95–132 (2017).

3

Reprogramming

3.1   Introduction

Since the first report on somatic reprogramming of human adult cells to iPSCs [1,2], rapid advances have enabled derivation from various somatic cells using different reprogramming methods from diverse patient populations, cultured with different media and matrices [3–7]. In order to enable clinical applications, there is a critical need to ensure the absence of detectable transgene delivering the reprogramming genes in the reprogrammed cells. The three most commonly used reprogramming methods to derive footprint free iPSC are non-integrating RNA virus such as Sendai [3], episomal DNA vectors [5], and mRNA [8]. A comparison of the three methods shows no significant difference in the overall quality of the resulting iPSC [9]. The CytoTune™-iPS 2.0 Sendai reprogramming kit is ideal for efficient and consistent generation of iPSCs from diverse genetic background and somatic cell sources. This chapter describes detailed methods for the generation of iPSCs under feeder-dependent and feeder-free conditions from fibroblasts and blood.

3.2   Materials

All materials are from Thermo Fisher Scientific unless specified otherwise.

Human Pluripotent Stem Cells: A Practical Guide, First Edition. Uma Lakshmipathy, Chad C. MacArthur, Mahalakshmi Sridharan and Rene H. Quintanilla.
© 2018 John Wiley & Sons, Inc. Published 2018 by John Wiley & Sons, Inc.

3.2.1 CytoTune-iPS 2.0 Sendai Reprogramming of Fibroblasts

1) CytoTune-iPS 2.0 Sendai Reprogramming Kit *Cat# A16517*
2) Human neonatal foreskin fibroblast cells (BJ Strain) *ATCC Cat# CRL2522*
3) DMEM (1×), Liquid (High Glucose) with GlutaMAX™-I *Cat# 10569-010*
4) Fetal Bovine Serum, ES Cell-Qualified *Cat # 16141-061*
5) DMEM/F-12 (1×), Liquid (1:1), with GlutaMAX™-I *Cat# 10565-018*
6) MEM Non-Essential Amino Acids Solution (100X) *Cat# 11140-050*
7) Knockout™ Serum Replacement, KSR *Cat# 10828-010*
8) 2-Mercaptoethanol (1000×), Liquid *Cat# 21985-023*
9) FGF-basic (AA 1-155) Recombinant Human *Cat# PHG0264*
10) Dulbecco's Phosphate Buffered Saline (DPBS) without calcium and magnesium *Cat# 14190-144*
11) TrypLE™ Express Cell Dissociation Reagent *Cat# 12604-013*
12) 0.05% Trypsin-EDTA (1×), Liquid *Cat# 25300-054*
13) Penicillin-Streptomycin, Liquid *Cat# 15140-122* (Optional)
14) BD PrecisionGlide Single Use Needle, 27 Gauge *Fisher Cat# 14-821-13B*
15) BD Disposable Syringes *Fisher Cat# 14-823-40*
16) Essential 8™ Medium *Cat# A1517001*
17) Geltrex™ LDEV-Free hESC-qualified Reduced Growth Factor Basement Membrane Matrix *Cat# A1413302*
18) Vitronectin (VTN-N) Recombinant Human Protein, Truncated *Cat# A14700*
19) UltraPure™ 0.5M EDTA, pH 8.0, *Cat# 15575020*
20) Versene Solution, *Cat# 15040066*

3.2.2 CytoTune 2.0 Reprogramming of CD34+

1) CytoTune-iPS 2.0 Sendai Reprogramming Kit *Cat# A16517*
2) StemPro-34 SFM Medium and CD34- cell kit *Cat# A14059*
3) GlutaMAX™ Supplement *Cat# 35050061*
4) SCF (C-Kit Ligand), Recombinant Human *Cat# PHC2111*
5) IL-3, Recombinant Human *Cat# PHC 0031*
6) GM-CSF, Recombinant Human *Cat# PHC2011*

7) Mouse (ICR) Inactivated Embryonic Fibroblasts *Cat# A24903*
8) Attachment Factor Protein (1×) *Cat#S-006-100*
9) DMEM/F-12 (1×), Liquid (1:1), with GlutaMAX™-I *Cat# 10565-018*
10) Fetal Bovine Serum (FBS) ES Cell-qualified *Cat# 16141-061*
11) KnockOut™ Serum Replacement (KSR) *Cat# 10828-010*
12) MEM Non-Essential Amino Acids *Cat# 11140-050*
13) Basic FGF, Recombinant Human *Cat# PHG0264*
14) 2-Mercaptoethanol (1000×), Liquid *Cat# 21985-023*
15) *Optional*: Penicillin-Streptomycin, Liquid *Cat#15140-122*
16) *Optional*: Polybrene Hexadimethrine Bromide *Sigma, Cat#H9268*
17) BD PrecisionGlide Single Use Needle, 27 Gauge *Fisher Cat# 14-821-13B*
18) BD Disposable Syringes *Fisher Cat# 14-823-40*
19) Dulbecco's Phosphate Buffered Saline (DPBS) without calcium and magnesium *Cat# 14190-144*
20) Essential 8™ Medium *Cat# A1517001*
21) Geltrex™ LDEV-Free hESC-qualified Reduced Growth Factor Basement Membrane Matrix *Cat# A1413302*
22) Vitronectin (VTN-N) Recombinant Human Protein, Truncated *Cat# A14700*
23) UltraPure™ 0.5M EDTA, pH 8.0, *Cat# 15575020*
24) Versene Solution, *Cat# 15040066*

3.2.3   CytoTune 2.0 Reprogramming of PBMC

1) CytoTune-iPS 2.0 Sendai Reprogramming Kit *Cat# A16517*
2) StemPro-34 SFM Medium and CD34- cell kit *Cat# A14059*
3) SCF (C-Kit Ligand), Recombinant Human *Cat# PHC2111*
4) IL-3, Recombinant Human *Cat# PHC 0031*
5) FLT3 Ligand Recombinant Human Protein *Cat# PHC9414*
6) IL6 Recombinant Human Protein *Cat# PHC0061*
7) Mouse (ICR) Inactivated Embryonic Fibroblasts *Cat# A24903*
8) Attachment Factor Protein (1×) *Cat# S-006-100*
9) DMEM/F-12 (1×), Liquid (1:1), with GlutaMAX™-I *Cat# 10565-018*
10) Fetal Bovine Serum (FBS) ES Cell-qualified *Cat# 16141-061*
11) KnockOut™ Serum Replacement (KSR) *Cat# 10828-010*
12) MEM Non-Essential Amino Acids *Cat# 11140-050*

13) Basic FGF, Recombinant Human *Cat# PHG0264*
14) 2-Mercaptoethanol (1000×), Liquid *Cat# 21985-023*
15) *Optional*: Penicillin-Streptomycin, Liquid *Cat# 15140-122*
16) *Optional*: Polybrene Hexadimethrine Bromide *Sigma, Cat# H9268*
17) BD PrecisionGlide Single Use Needle, 27 Gauge *Fisher Cat# 14-821-13B*
18) BD Disposable Syringes *Fisher Cat# 14-823-40*
19) Dulbecco's Phosphate Buffered Saline (DPBS) without calcium and magnesium *Cat# 14190-144*
20) Essential 8™ Medium *Cat# A1517001*
21) Geltrex™ LDEV-Free hESC-qualified Reduced Growth Factor Basement Membrane Matrix *Cat# A1413302*
22) Vitronectin (VTN-N) Recombinant Human Protein, Truncated *Cat# A14700*
23) UltraPure™ 0.5M EDTA, pH 8.0, *Cat# 15575020*
24) Versene Solution, *Cat# 15040066*

3.2.4   RT-PCR for SEV Primer Detection

1) TRIzol™ *Cat # 5596-026*
2) Chloroform *Sigma Cat# C-2432*
3) Isopropanol *Sigma Cat#I9516-500ml*
4) Ethanol *Sigma Cat#E7023-500ml*
5) RNase-free Water *Cat#10977*
6) DNA-*free*™ Kit *Cat#AM1906*
7) High-capacity cDNA Reverse Transcription kit with RNase Inhibitor, 200 reactions *Cat#4374966*
8) SuperScript™ VILO cDNA Synthesis Kit *Cat#11754-050*
9) AccuPrime™ SuperMix *Cat#12342-010*

3.2.5   Epi 5 Reprogramming of Fibroblasts

1) Epi5™ Episomal iPSC Reprogramming Kit *Cat# A15960*
2) Human neonatal foreskin fibroblast cells (strain BJ) *ATCC no. CRL2522*
3) Dulbecco's Modified Eagle Medium (DMEM) with GlutaMAX™-I (high glucose) *Cat# 10569*
4) KnockOut™ DMEM/F-12 *Cat# 12660-012*
5) Fetal Bovine Serum (FBS), ESC-Qualified *Cat# 16141-079*

6) MEM Non-essential Amino Acids (NEAA) *Cat# 11140-050*
7) Basic Fibroblast Growth Factor (bFGF), recombinant human *Cat#. PHG0264*
8) Essential 8™ Medium, consisting of Essential 8™ Basal Medium and Essential 8™ Supplement (50×) *Cat# A1517001*
9) N-2 Supplement (100×) *Cat# 17502-048*
10) B-27™ Serum-Free Supplement (50×) *Cat# 17504-044*
11) GlutaMAX™-I Supplement *Cat# 35050-061*
12) β-mercaptoethanol *Cat# 21985-023*
13) Geltrex™ LDEV-Free hESC-Qualified Reduced Growth Factor BasementMembrane Matrix *Cat# A1413301*
14) 0.05% Trypsin-EDTA (1×), Phenol Red *Cat# 25300-054*
15) UltraPure™ 0.5 M EDTA, pH 8.0 *Cat# 15575-020*
16) Dulbecco's PBS (DPBS) without Calcium and Magnesium *Cat# 14190144*

3.2.6   Epi 5 Reprogramming of CD34+

1) Epi5™ Episomal iPSC Reprogramming Kit *Cat# A15960*
2) StemPro-34 SFM Medium and CD34- cell kit *Cat# A14059*
3) GlutaMAX™ Supplement *Cat# 35050061*
4) SCF (C-Kit Ligand), Recombinant Human *Cat# PHC2111*
5) IL-3, Recombinant Human *Cat# PHC 0031*
6) GM-CSF, Recombinant Human *Cat# PHC2011*
7) Mouse (ICR) Inactivated Embryonic Fibroblasts *Cat# A24903*
8) Attachment Factor Protein (1×) *Cat#S-006-100*
9) DMEM/F-12 (1×), Liquid (1:1), with GlutaMAX™-I *Cat# 10565-018*
10) Fetal Bovine Serum (FBS) ES Cell-qualified *Cat# 16141-061*
11) KnockOut™ Serum Replacement (KSR) *Cat# 10828-010*
12) MEM Non-Essential Amino Acids *Cat# 11140-050*
13) Basic FGF, Recombinant Human *Cat# PHG0264*
14) 2-Mercaptoethanol (1000×), Liquid *Cat# 21985-023*
15) *Optional*: Penicillin-Streptomycin, Liquid *Cat#15140-122*
16) *Optional*: Polybrene Hexadimethrine Bromide *Sigma, Cat#H9268*
17) BD PrecisionGlide Single Use Needle, 27 Gauge *Fisher Cat# 14-821-13B*
18) BD Disposable Syringes *Fisher Cat# 14-823-40*

19) Dulbecco's Phosphate Buffered Saline (DPBS) without calcium and magnesium *Cat# 14190-144*
20) Essential 8™ Medium *Cat# A1517001*
21) Geltrex™ LDEV-Free hESC-qualified Reduced Growth Factor Basement Membrane Matrix *Cat# A1413302*
22) Vitronectin (VTN-N) Recombinant Human Protein, Truncated *Cat# A14700*
23) UltraPure™ 0.5M EDTA, pH 8.0, *Cat# 15575020*

3.2.7   PCR Detection of Epi5 Vectors

1) CellsDirect Resuspension & Lysis Buffers *Cat# 11739-010*
2) AccuPrime Taq DNA Polymerase High Fidelity *Cat# 12346-086*

3.2.8   Terminal AP Staining

1) UltraPure 1M Tris-HCL, pH 8.0
2) Distilled Water
3) VECTOR Red Alkaline Phosphatase (Red AP) Substrate Kit (Vector Labs *Cat# SK-5100*)

3.3   Solutions

3.3.1   Fibroblast Medium (for 500 mL)

DMEM	444.5 mL
FBS (ES qualified)	50 mL
NEAA	5 mL

Sterilize through 0.22 µm filter. Medium lasts for up to 1 month at 4 °C.

3.3.2   N-2/B27 Medium (500 mL)

DMEM/F-12	479 mL
N-2 Supplement (100×)	5 mL
B-27 Supplement (50×)	10 mL
MEM NEAA (10 mM)	5 mL
2-Mercaptoethanol (55 mM)	908 µL

Prepare the N2/B27 medium without bFGF, and then supplement with fresh bFGF to a final concentration of 100 ng/mL when the medium is used.

N2/B27 medium (without bFGF) can be stored at 2–8 °C for up to 1 week.

3.3.3   PBMC Media (for 500 mL)

StemPro-34 Medium	482 mL
StemPro-34 Nutrient Supplement	13 mL
GlutaMAX-I (100×)	5 mL

Before use, add the following cytokines to the indicated final concentration into the medium. These must be added freshly, just prior to use.

SCF	100 ng/mL
FLT3	100 ng/mL
IL3	20 ng/mL
IL6	20 ng/mL

3.3.4   CD34+ Media (500 mL)

StemPro-34 Medium	482 mL
StemPro-34 Nutrient Supplement	13 mL
GlutaMAX-I (100×)	5 mL

Before use, add the following cytokines to the indicated final concentration into the medium. These must be added freshly, just prior to use.

SCF	100 ng/mL
FLT3	50 ng/mL
GM-CSF	25 ng/mL

3.3.5   iMEF Medium (for 500 mL) for Feeder-Dependent Generation of iPSCs

DMEM	444.5 mL
FBS (ES qualified)	50 mL
NEAA	5 mL
2-Mercaptoethanol	500 μL

Sterilize through 0.22 μm filter. Medium lasts for up to 1 month at 4 °C.

3.3.6   Human PSC Medium (for 500 mL) for Feeder-Dependent Generation of iPSCs

DMEM-F12	395 mL
KnockOut Serum Replacement	100 mL
NEAA	5 mL
2-Mercaptoethanol	500 μL

Sterilize through 0.22 μm filter. Medium lasts for up to 1 month at 4 °C.

Add bFGF (final concentration 4 ng/mL) fresh prior to use (example: 0.4 μL reconstituted bFGF per mL of medium).

3.3.7   Basic FGF Solution (10 μg/mL, for 1 mL)

Basic FGF	10 μg
D-PBS (-/-)	990 μL
KnockOut Serum Replacement	10 μL

Sterilize through 0.22 μm filter. Medium lasts for up to 1 month at 4 °C.

Aliquot and store at −20 °C for up to 3 months. Once bFGF aliquot is thawed, use within 7 days, when stored at 4 °C.

3.3.8   Essential 8 Medium (for 500 mL) for Feeder-Free Generation of iPSCs

Essential 8 Basal Medium	490 mL
Essential 8 Supplement	10 mL

Sterilize through 0.22 μm filter. Thaw the frozen Essential 8 supplement at 2–8 °C overnight or room temperature for 2 hours. Do not thaw the frozen supplement at 37 °C. Store Essential 8 medium in a polystyrene bottle at 2–8 °C for up to 2 weeks. Before use, warm complete medium required for that day at room temperature until it is no longer cool to the touch. Do not warm the medium at 37 °C.

3.3.9   200 mM Tris-HCL Solution (pH 8.0) (for 500 mL)

UltraPure 1M Tris-HCL, pH 8.0	100 mL
Distilled Water	400 mL

The diluted Tris solution can be stored at room temperature for up to 4 weeks.

3.4   Methods

3.4.1   CytoTune Reprogramming

The CytoTune-iPS 2.0 reprogramming system uses vectors based on a modified, non-transmissible form of the Sendai virus (SeV) to safely and effectively deliver and express key genetic factors necessary for reprogramming somatic cells into iPSCs. In contrast to many available protocols, which rely on viral vectors that integrate into the genome of the host cell, the CytoTune-iPS 2.0 reprogramming system uses vectors that are non-integrating and remain in the cytoplasm (i.e., they are zero-footprint). In addition, the host cell can be cleared of the vectors and exogenous reprogramming genes by exploiting the cytoplasmic nature of SeV and the functional temperature sensitivity mutations introduced into the key viral proteins (Figure 3.1).

Reprogramming is a multistep protocol comprising two main stages (Figure 3.2). The first stage consists of somatic cell manipulation. This includes culture of somatic cells such as fibroblasts or blood, transduction with CytoTune, recovery, and replating onto either feeder or feeder-free PSC matrices (e.g., iMEF or VTN). The second stage consists of culture of the plated cells in PSC media to facilitate the emergence of iPSC colonies.

SeV(PM)**KOS**/TS12ΔF

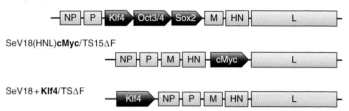

SeV18(HNL)**cMyc**/TS15ΔF

SeV18 + **Klf4**/TSΔF

Figure 3.1 The CytoTune-iPS 2.0 Sendai reprogramming kit contains three SeV-based reprogramming vectors that have been optimized for generating iPSC from human somatic cells; the first vector contains KOS (KLF4, OCT4, and SOX2); the second vector contains c-MYC; and the third vector contains additional KLF4 to achieve higher reprogramming efficiency.

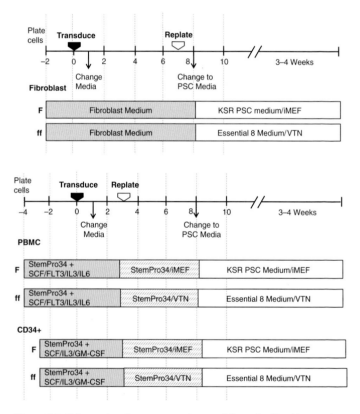

Figure 3.2 Schematics of reprogramming workflows for fibroblast and blood cells in feeder-dependent (F) and feeder-free conditions (ff).

3.4.1.1   Reprogramming Human Fibroblasts with CytoTune

The use of BJ fibroblasts serves as a model of the typical reprogramming protocol, which will need to be modified for different starting populations of somatic cells. It is important that the starting population of the cells be reprogrammed at a very early passage (P2–P6) and in the log phase of replication for optimal reprogramming. The following protocol describes the procedures for fibroblast growth directly from cryopreserved stocks.

3.4.1.1.1   Day -7: Thaw Fibroblasts

1) Prewarm the fibroblast medium in a 37 °C water bath.
2) Wear eye protection as cryogenic vials stored in the liquid phase of liquid nitrogen may accidentally explode when warmed.
3) Wear ultra-low temperature cryogenic gloves when removing the vial from cryogenic storage. Remove the cryogenic vial of BJ fibroblasts from the liquid nitrogen storage tank using metal forceps.
4) Roll the vial between your hands for about 10–15 seconds to remove frost.
5) Immerse the vial in a 37 °C water bath. Swirl gently. *Note*: do not submerge the cap to prevent contamination.
6) When only an ice crystal remains (2–3 minutes), remove the vial from the water bath.
7) Spray the outside of the vial with 70% ethanol and place in the culture hood.
8) Pipette cells into a 50 mL conical tube with a 1 mL pipette (or P1000).
9) Using a 10 mL pipette, add 10 mL of prewarmed fibroblast medium to the 50 mL conical tube drop-wise (to avoid osmotic shock to cells), while gently swirling the conical tube. Pipette up and down gently to mix the cell suspension. *Note*: this is important to properly dilute out the cryo-protectant.
10) Transfer the cell suspension to a 15 mL conical tube and centrifuge at 200× g for 2 minutes.
11) Gently aspirate off the supernatant, without disturbing the cell pellet.

12) Reconstitute the cell pellet in the appropriate amount of fibroblast medium. Typically seed the contents of the vial in 15 mL of medium in a T-75 TC treated culture flask.
13) Incubate the fibroblast culture in a 37 °C, 5% CO_2 incubator overnight.
14) The following day, aspirate off the spent medium and replace with fresh, prewarmed fibroblast medium.
15) Change the medium every other day until the flask is approximately 70% confluent (usually 4–5 days).
16) When the fibroblasts are approximately 70% confluent, the cultures must be split and reseeded for creating cell banks, or seeded directly for reprogramming experiments at the appropriate density for CytoTune transductions. It is recommended to use a portion of the early passage cells for experimentation.
17) To split fibroblast cultures, remove the FBS containing fibroblast media from the culture vessel via gentle aspiration.
18) Wash cells twice with 5 mL of room temperature D-PBS without calcium or magnesium. *Note*: divalent cations must not be present for proper disassociation of the fibroblasts from the culture vessel.
19) Aspirate off the D-PBS wash and add 3 mL of 0.05% Trypsin-EDTA (or TrypLE Express), at room temperature, to the culture vessel (adjust volume of Trypsin up or down for appropriate culture vessel size).
20) Incubate the culture in a 37 °C, 5% CO_2 incubator for approximately 3–5 minutes. Observe your cells under a microscope and stop the incubation when the cells start to round up.
21) Add 6 mL of prewarmed fibroblast media to the culture vessel to stop the trypsinization. *Note*: FBS contains endogenous trypsin inhibitors, which stop further enzymatic action.
22) Wash off all the cells from the vessel surface by flushing the suspension several times to harvest all the cells in a single cell suspension. Transfer the cell suspension to a 15 mL centrifuge tube.
23) Centrifuge the cell suspension at 200× g for 2 minutes.
24) Following centrifugation, gently aspirate off the supernatant, without disturbing the cell pellet.

25) Resuspend the cell pellet with 2–3 mL of fresh prewarmed fibroblast media and pipette up and down to create a single cell suspension.

26) Prepare a cell count and seed the appropriate amount of cells into a new TC treated culture dish with the appropriate volume of media. Refer to Table 2.2 for recommended volumes for each type of vessel. Typically a 1 to 6 split can be used for regular maintenance of the fibroblasts, generally 3–4 days between passages.

3.4.1.1.2 Day -2: Plate Fibroblasts for Transduction

27) Two days before transduction, plate fibroblast cells into at least two wells of a six-well plate at an appropriate density to achieve 2×10^5 to 3×10^5 cells per well on the day of transduction, Day 0 (Figure 3.3). One of the wells will be used to count cells for viral volume calculations and will not be transduced. *Note*: we recommend about 50–60% confluency on the day of transduction, as overconfluency results in decreased reprogramming efficiency. It is recommended that the entire CytoTune 2.0 kit is used when it is initially thawed. Refreezing and thawing of the virus will result in lower titer, and is not recommended. As such, it is suggested that experiments are planned so that sufficient reactions are set up in order to use the entire kit.

28) Culture the cells for two more days, ensuring the cells have fully adhered and extended.

(a) (b)

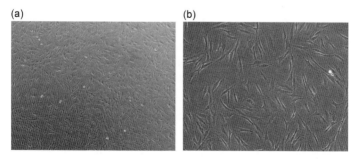

Figure 3.3 BJ fibroblasts seeded at desired densities, at 50× magnification (a) and 100× magnification (b).

3.4.1.1.3 Day 0: Perform the Transductions

29) Harvest the cells from one well to perform a cell count as follows. *Note*: these cells will not be transduced, but will be used to estimate the cell number in the other well(s) plated in on Day −2.
    a) Wash the well once with D-PBS (-/-) for 2–3 minutes at room temperature.
    b) Aspirate off the D-PBS wash and add 0.5 mL of room temperature 0.05% Trypsin-EDTA (or TrypLE Express), to the culture vessel.
    c) Incubate the culture in a 37 °C, 5% CO_2 incubator for approximately 3–5 minutes. Observe your cells under a microscope and stop incubation when the cells start to round up.
    d) When the cells have rounded up, add 0.5 mL of fibroblast medium into the well, and collect the cells in a 15 mL conical centrifuge tube.
30) Count the cells using the desired method (e.g., Countess automated cell counter or hemacytometer), and calculate the volume of each virus needed to reach the target MOI using the live cell count and the titer information on the CoA (www.thermofisher.com/cytotune), following the equation in Table 3.1. *Note*: we recommend initially performing the transductions with MOIs of 5, 5, and 3 (i.e., KOS MOI = 5, hc-Myc MOI = 5, hKlf4 MOI = 3). These MOIs can be optimized for your application.
31) For each well to be transduced, prewarm 1 mL of fibroblast in a 37 °C water bath.
32) Remove one set of CytoTune 2.0 Sendai tubes from the −80 °C storage. Remove the vials from the pouch and thaw each tube one at a time by first immersing the bottom of the tube in a 37 °C water bath for 5–10 seconds, and then removing the tube from the water bath and allowing it to

Table 3.1 Formula to calculate the volume of each virus needed to reach the target MOI(s).

$$\text{Volume of virus } (\mu L) = \frac{\text{MOI (CIU/cell)} \times \text{number of cells}}{\text{Titer of virus (CIU/mL)} \times 10^{-3} \text{ (mL/}\mu L)}$$

thaw at room temperature. Once thawed, briefly centrifuge the tube and place it immediately on ice.

33) Add the calculated volumes of each of the 3 CytoTune 2.0 Sendai tubes to the prewarmed fibroblast medium.

34) Ensure that the solution is thoroughly mixed by pipetting the mixture gently up and down. Complete the next step within 5 minutes.

35) Aspirate the fibroblast medium from the well(s) to be transduced, and add 1 mL of the solution prepared in Step 33 to each well(s). Place the cells in a 37 °C, 5% CO_2 incubator and incubate overnight.

3.4.1.1.4   Day 1: Remove the Virus

36) Following overnight incubation, aspirate off the transduction medium and dispose of it properly. Replace with 2 mL of prewarmed, fresh fibroblast medium per well, and continue to culture the transduced cells. *Note*: depending on your cell type, you should expect to see some cytotoxicity 24–48 hours post transduction, which can affect >50% of your cells. This is an indication of high uptake of the virus. We recommend that you continue culturing your cells and proceed with the protocol.

3.4.1.1.5   Day 2 to Day 6: Feed Cells

37) Culture the cells for 6 more days, changing the spent medium with fresh fibroblast medium every other day. *Note*: depending on your cell type, you may observe high cell density before Day 5. We do *not* recommend passaging your cells onto iMEF culture dishes or Geltrex or vitronectin coated dishes before 7 days post transduction.

38) Next, proceed with Option A *or* Option B.

3.4.1.1.6   Option A: Feeder-Dependent iPSC Generation
Day 6: Prepare iMEF Coated Culture Dishes

1) Prepare iMEF coated culture dishes as previously described. Typically six-well dishes are used for iPSC generation.

Day 7: Harvest and Replate Transduced Cells

2) Seven days after transduction, the cells are ready to be harvested and plated on iMEF culture dishes. Aspirate off the

normal culture medium from the transduced fibroblasts, and wash the cells once with 2 mL of D-PBS (-/-).

3) Aspirate off the D-PBS wash and add 0.5 mL of TrypLE Select reagent or 0.05% trypsin/EDTA. Incubate the cells for 2–5 minutes at 37 °C. *Note*: because the cells can be very sensitive to trypsin at this point, minimize trypsin exposure time.

4) When the cells have rounded up, add 2 mL of fibroblast medium into each well to stop the trypsinization.

5) Dislodge the cells from the well by gently pipetting the media across the surface of the dish. Collect the cell suspension in a 15 mL conical centrifuge tube.

6) Centrifuge the cells at 200× g for 4 minutes.

7) Aspirate off the supernatant, and resuspend the cells in an appropriate amount of fresh, prewarmed fibroblast medium.

8) Count the cells using the desired method (e.g., Countess automated cell counter or hemacytometer), and seed the iMEF culture dishes with 2.5×10^4 to 2×10^5 cells per well of a six-well plate, and incubate in a 37 °C, 5% CO_2 incubator overnight. It is recommended to plate at least two to three different densities, depending on your cell type, to ensure sufficient numbers of colonies. *Note*: if there are leftover cells, they should be used as a positive control when performing RT-PCR or qPCR to determine presence of the Sendai virus in the resulting iPSCs. Freeze a cell pellet at −80 °C, or resuspend in 0.5 mL of TRIzol and store at −80°C.

Day 8 to Day 21: Feed Cells with iPSC Medium

9) Twenty-four hours after reseeding the transduced cells on iMEF coated vessels, change the fibroblast medium to iPSC medium.

10) Replace the spent iPSC medium every day thereafter.

11) Starting on Day 8 post transduction, observe the plates every other day under a microscope for the emergence of cell clumps indicative of reprogrammed cells (Figure 3.4). *Note*: for BJ fibroblasts, we normally observe initial colony formation on Day 12 post transduction. However, depending on your cell type, you may need to culture for up to 3–4 weeks before seeing colonies.

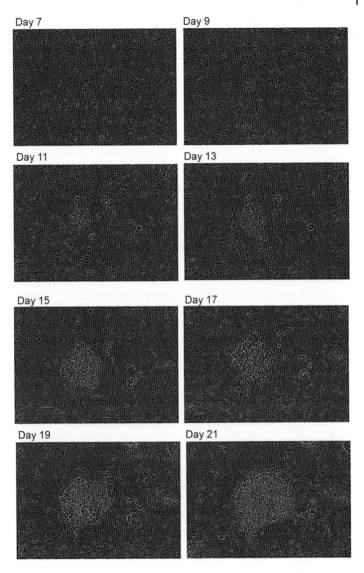

Figure 3.4 Time course of fibroblast reprogramming under feeder-dependent conditions. 50× phase contrast images: after seeding CytoTune transduced fibroblasts onto iMEF-coated dishes (day 7), and images of emerging iPSC colonies from day 9 to day 21.

12) By day 15–21 after transduction, colonies should have grown to an appropriate size for transfer. The day before transferring the colonies, prepare iMEF coated culture six-well plates. *Note*: we typically harvest colonies closer to 21 days or less, to avoid differentiation.

13) When colonies are ready for transfer, perform live staining using Tra1-60 or Tra1-81 for selecting reprogrammed colonies if desired.

14) Manually pick colonies and transfer them onto prepared iMEF plates.

3.4.1.1.7   Option B: Feeder-Free Generation of iPSC Colonies
Day 7: Harvest and Replate Transduced Cells

1) Prepare VTN or Geltrex coated culture dishes as described previously. Typically six-well dishes are used for iPSC generation.

2) Seven days after transduction, the cells are ready to be harvested and plated on culture dishes. Aspirate off the normal culture medium from the transduced fibroblasts, and wash the cells once with 2 mL of D-PBS (-/-).

3) Aspirate off the D-PBS wash and add 0.5 mL of TrypLE Select reagent or 0.05% trypsin/EDTA. Incubate the cells for 2–5 minutes at 37 °C. *Note*: because the cells can be very sensitive to trypsin at this point, minimize trypsin exposure time.

4) When the cells have rounded up, add 2 mL of fibroblast medium into each well to stop the trypsinization.

5) Dislodge the cells from the well by gently pipetting the media across the surface of the dish. Collect the cell suspension in a 15 mL conical centrifuge tube.

6) Centrifuge the cells at 200× g for 4 minutes.

7) Aspirate off the supernatant, and resuspend the cells in an appropriate amount of fresh, prewarmed fibroblast medium.

8) Count the cells using the desired method (e.g., Countess automated cell counter or hemacytometer), and seed the culture dishes with 2.5×10^4 to 2×10^5 cells per well of a six-well plate, and incubate in a 37 °C, 5% CO_2 incubator overnight. It is recommended to plate at least two to three different densities, depending on your cell type, to ensure sufficient numbers of colonies. *Note*: if there are leftover cells, they should be used as a positive control when performing RT-PCR or

qPCR to determine presence of the Sendai virus in the resulting iPSCs. Freeze a cell pellet at −80 °C or resuspend in 0.5 mL of TRIzol and store at −80 °C.

Day 8 to Day 21: Feed Cells with Essential 8 Medium

9) Twenty-four hours after reseeding on the desired matrix, change the fibroblast medium to Essential 8 medium, and replace the spent medium with fresh Essential 8 medium every day thereafter.

10) Starting on Day 8, observe the plates every other day under a microscope for the emergence of cell clumps indicative of reprogrammed cells (Figure 3.5). *Note*: for BJ fibroblasts, we normally observe initial colony formation on Day 12 post transduction. However, depending on your cell type, you may need to culture for up to 3–4 weeks before seeing colonies.

11) By day 15–21 after transduction, colonies should have grown to an appropriate size for transfer. The day of transferring the colonies, prepare vitronectin or Geltrex coated culture six-well plates. *Note*: we typically harvest colonies closer to 21 days or less, to avoid differentiation.

12) When colonies are ready for transfer, perform live staining using Tra1-60 or Tra1-81 for selecting reprogrammed colonies if desired.

13) Manually pick colonies and transfer them onto prepared culture plates.

3.4.1.2 Reprogramming Peripheral Blood Mononucleocyte Cells with CytoTune

The following protocol has been optimized for peripheral blood mononuclear cells (PBMCs) isolated through density gradient centrifugation via Ficoll-Paque. We recommend that you optimize the protocol for your cell type and media conditions.

3.4.1.2.1 Day −4: Seed PBMCs

1) Four days before transduction, remove vial(s) of PBMCs from liquid nitrogen storage.

2) Thaw the vial quickly in a 37 °C water bath. When only a small ice crystal remains in the vial, remove it from the water bath (2–3 minutes). Spray the outside of the vial with 70% ethanol before placing it in the cell culture hood.

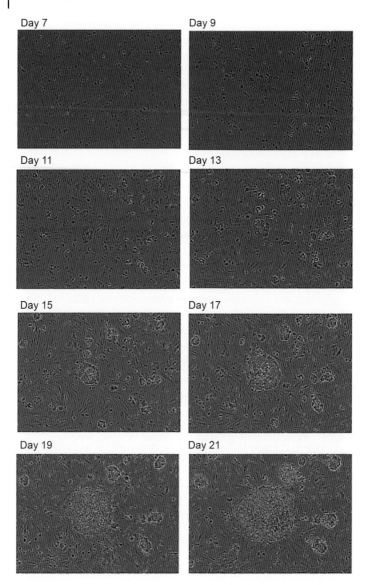

Figure 3.5 Time course of fibroblast reprogramming under feeder-free conditions. 50× phase contrast images: after seeding CytoTune transduced fibroblasts onto Geltrex-coated dishes (day 7) and images of emerging iPSC colonies from day 9 to day 21.

3) Gently transfer the PBMCs into a 15 mL conical tube. Slowly (drop-wise) add 5–10 mL prewarmed complete PBMC medium to the cell suspension. Remove an aliquot of cells to count and determine cell viability. *Note*: PBMC medium consists of complete StemPro-34 medium containing the appropriate cytokines; aliquot the cytokines and add fresh daily.

4) Centrifuge the cell suspension at 200× g for 10 minutes.

5) Gently discard the supernatant, and resuspend the cells in complete PBMC medium to a density of 5×10^5 cells/mL, using the appropriate amount of medium based on the cell count.

6) Add 1 mL of the cell suspension per well to the middle section of a 24-well plate to prevent excessive evaporation of the medium during incubation.

7) Incubate the cells in a 37 °C incubator with a humidified atmosphere of 5% CO_2.

3.4.1.2.2   *Day –3 to Day –1: Observe Cells and Add Fresh Medium*

8) Gently remove 0.5 mL of the medium from each well without disturbing the cells at the bottom of the well.

9) Add 0.5 mL of fresh complete PBMC medium to each well, trying not to disturb the cells. If cells are present in the 0.5 mL removed from the wells, centrifuge the cell suspension at 200× g for 10 minutes, discard the supernatant and resuspend the cells in 0.5 mL fresh PBMC medium before adding them back to the plate. *Note*: some cell death is generally observed the first day after the thaw. Some cells may adhere to the surface of the tissue culture plate. Proceed with the cells in suspension. Cells may not proliferate, but should maintain stable cell number for the first few days (PBMCs contain a variety of cells, and the current media system is only targeting a small population).

3.4.1.2.3   *Day 0: Perform the Transductions*

10) For each transduction reaction, prewarm 1 mL of PBMC medium in a 37 °C water bath.

11) Count the cells using the desired method (e.g., Countess automated cell counter), and calculate the volume of each

virus needed to reach the target MOI using the *live cell count* and the titer information on the CoA (www.thermofisher.com/cytotune), following the equation in Table 3.1. We recommend 2.5×10^5 to 5×10^5 cells per transduction. *Note*: we recommend initially performing the transductions with MOIs of 5, 5, and 3 (i.e., KOS MOI = 5, hc-Myc MOI = 5, hKlf4 MOI = 3). These MOIs can be optimized for your application.

12) Add the appropriate volume of cells to achieve the desired cell number to a sterile, round-bottom culture tube.

13) Remove CytoTune 2.0 Sendai tubes from the −80 °C storage. Thaw each tube one at a time by first immersing the bottom of the tube in a 37 °C water bath for 5–10 seconds, and then removing the tube from the water bath and allowing it to thaw at room temperature. Once thawed, briefly centrifuge the tube and place it immediately on ice.

14) Add the calculated volumes of each of the three CytoTune 2.0 Sendai tubes to 1 mL of PBMC medium, prewarmed to 37 °C. Ensure that the solution is thoroughly mixed by pipetting the mixture gently up and down. Complete the next step within 5 minutes.

15) Add 1 mL of the virus solution prepared in Step 14 above to each tube(s) of cells prepared in step 12 above.

16) Centrifuge the cells and virus at 1000× g for 30 minutes at room temperature.

17) Transfer the cells and the medium containing the virus to a 12-well plate, and add additional PBMC medium to bring the total volume to 2 mL. Place the cells in a 37 °C, 5% CO_2 incubator and incubate overnight. *Note*: if preferred, this centrifugation step can be performed in a 12-well plate. Add the desired number of cells to one well, and then add the virus solution to the cells. Seal the edges of the plate with Parafilm laboratory film and centrifuge at 1000× g for 90 minutes at room temperature. Add an additional 1 mL of complete PBMC medium to each well and incubate the plate overnight at 37 °C in a humidified atmosphere of 5% CO_2. Although this centrifugation step is not essential, it significantly increases the transduction and reprogramming efficiencies. Centrifugation in a round-bottom tube

will yield the best results, while centrifugation in a 12-well plate will yield slightly lower efficiencies, and no centrifugation will yield much lower efficiencies. If the centrifugation step is omitted, transductions can be performed in a 24-well plate using 0.3 mL of total volume of cells, virus, and medium. Adding 4 μg/mL of Polybrene to the medium at the time of transduction may increase transduction efficiencies only if the centrifugation step is not performed.

3.4.1.2.4 Day 1 to Day 3: Remove the Virus and Culture the Cells

18) The next day, remove the cells and medium from the culture plate and transfer to a 15 mL centrifuge tube. Rinse the well gently with 1 mL of medium to ensure most of the cells are harvested.
19) Remove the CytoTune 2.0 Sendai viruses by centrifuging the cell suspension at 200× g for 10 minutes.
20) Gently aspirate off the supernatant, and resuspend the cells in 0.5 mL of complete PBMC medium per well of a 24-well plate. *Note*: the cells may have drastic cell death (>60%); continue with the protocol. For the first 48 hours, observe the cells under the microscope for changes in cell morphology as a validation of transduction. Expect large, aggregated cells.
21) Culture the cells at 37 °C in a humidified atmosphere of 5% CO_2 for 2 days.
22) Next, proceed with Option A or Option B.

3.4.1.2.5 Option A: Feeder-Dependent Reprogramming
Day 2: Prepare iMEF Culture Dishes

1) Prepare iMEF culture dishes as described previously. Typically six-well dishes are used for iPSC generation.

Day 3: Plate Transduced Cells on iMEF Culture Dishes

2) Count the cells using the desired method (e.g., Countess automated cell counter) and seed the six-well iMEF culture plates with 10 000 to 50 000 *live* cells per well in 2 mL of complete StemPro-34 medium (without cytokines). Plate any excess cells in additional iMEF culture dishes. *Note*: if there

are leftover cells, they should be used as a positive control when performing RT-PCR or qPCR to determine presence of Sendai virus in the resulting iPSCs. Freeze a cell pellet at $-80\,°C$, or resuspend in 0.5 mL of TRIzol and store at $-80\,°C$.

3) Incubate the cells at $37\,°C$ in a humidified atmosphere of 5% CO_2.

Day 4 to Day 6: Replace Spent Medium

4) Every other day, gently remove 1 mL (half) of the spent medium from the cells and replace it with 1 mL of fresh complete StemPro-34 medium (without cytokines), being careful not to disturb the cells. *Note*: if cells are present in the 0.5 mL removed from the wells, centrifuge the cell suspension at $200\times$ g for 10 minutes, discard the supernatant, and resuspend the cells in 0.5 mL fresh StemPro34 medium (without cytokines) before adding them back to the plate.

Day 7: Start Transitioning Cells to iPSC Medium

5) Remove 1 mL (half) of StemPro-34 medium (without cytokines) from the cells and replace it with 1 mL of iPSC medium to start the adaptation of the cells to the new culture medium.

Day 8 to Day 21: Feed and Monitor the Cells

6) Twenty-four hours later (day 8), change the full volume of the medium to iPSC medium, and replace the spent iPSC medium every day thereafter.

7) Starting on day 8, observe the plates every other day under a microscope for the emergence of cell clumps indicative of reprogrammed cells

8) By day 15–21 after transduction, colonies should have grown to an appropriate size for transfer. The day before transferring the colonies, prepare iMEF coated culture six-well plates. *Note*: we typically harvest colonies closer to 21 days or less, to avoid differentiation.

9) When colonies are ready for transfer, perform live staining using Tra1-60 or Tra1-81 for selecting reprogrammed colonies if desired.

10) Manually pick colonies and transfer them onto prepared iMEF plates.

3.4.1.2.6 Option B: Feeder-Independent Reprogramming
Day 3: Plate Transduced Cells

1) Prepare VTN or Geltrex coated culture dishes as described previously. Typically six-well dishes are used for iPSC generation.
2) Count the cells using the desired method (e.g., Countess automated cell counter) and seed the six-well VTN or Geltrex coated culture plates with 10 000 to 50 000 live cells per well in 2 mL of complete StemPro-34 medium (without cytokines). Plate any excess cells in an additional culture dishes. *Note*: if there are leftover cells, they should be used as a positive control when performing RT-PCR or qPCR to determine presence of Sendai virus in the resulting iPSCs. Freeze a cell pellet at −80 °C, or resuspend in 0.5 mL of TRIzol and store at −80 °C.
3) Incubate the cells at 37 °C in a humidified atmosphere of 5% CO_2.

Day 4 to Day 6: Replace Spent Medium

4) Every other day, gently remove 1 mL (half) of the spent medium from the cells and replace it with 1 mL of fresh complete StemPro-34 medium (without cytokines), being careful not to disturb the cells. *Note*: if cells are present in the 0.5 mL removed from the wells, centrifuge the cell suspension at 200× g for 10 minutes, discard the supernatant, and resuspend the cells in 0.5 mL fresh StemPro34 medium (without cytokines) before adding them back to the plate.

Day 7: Start Transitioning Cells to iPSC Medium

5) Remove 1 mL (half) of StemPro-34 medium from the cells and replace it with 1 mL of Essential 8 medium to start the adaptation of the cells to the new culture medium.

Day 8 to Day 21: Feed and Monitor the Cells

6) Twenty-four hours later (day 8), change the full volume of the medium to Essential 8 medium, and replace the spent Essential 8 medium every day thereafter.
7) Starting on day 8, observe the plates every other day under a microscope for the emergence of cell clumps indicative of reprogrammed cells.

8) By day 15–21 after transduction, colonies should have grown to an appropriate size for transfer. The day of transferring the colonies, prepare vitronectin or Geltrex coated culture six-well plates. *Note*: we typically harvest colonies closer to 21 days or less, to avoid differentiation.

9) When colonies are ready for transfer, perform live staining using Tra1-60 or Tra1-81 for selecting reprogrammed colonies if desired.

10) Manually pick colonies and transfer them onto prepared culture plates.

3.4.1.3 Reprogramming CD34+ Cells with CytoTune

The following protocol has been optimized for CD34+ cells derived from the human umbilical cord blood of mixed donors.

3.4.1.3.1 Day –3: Seed Cells

1) Three days before transduction, remove one vial of CD34+ cells from the liquid nitrogen storage tank.

2) Briefly roll the cryovial between hands to remove frost, and swirl it gently in a 37 °C water bath to thaw the StemPro CD34$^+$ cells.

3) When only a small ice crystal remains in the vial, remove it from water bath. Spray the outside of the vial with 70% ethanol before placing it in the cell culture hood.

4) Pipette the thawed cells gently into a 15 mL conical tube.

5) Add 10 mL of prewarmed complete StemPro-34 medium drop-wise to the cells. Gently mix by pipetting up and down. *Note*: adding the medium slowly helps the cells to avoid osmotic shock.

6) Centrifuge the cell suspension at 200× g for 10 minutes.

7) Gently discard the supernatant and resuspend the cells to a density of 5×10^5 cells/mL, using the appropriate amount of StemPro-34 medium containing cytokines (i.e., SCF, IL3, and GM-CSF).

8) Place 0.5 mL of cell suspension into each well of a 24-well plate and incubate at 37 °C in a humidified atmosphere of 5% CO_2. *Note*: we recommend using the wells in the middle section of the 24-well plate to prevent excessive evaporation of the medium during incubation.

3.4.1.3.2 Day −2: Observe Cells and Add Fresh Medium

9) Two days before transduction, add 0.5 mL of fresh complete StemPro-34 medium containing cytokines to each well, without disturbing the cells.

3.4.1.3.3 Day −1: Observe Cells and Add Fresh Medium

10) One day before transduction, gently remove 0.5 mL of medium and add 1 mL of fresh complete StemPro 34 medium containing cytokines without disturbing the cells. *Note*: if cells are present in the 0.5 mL removed from the wells, centrifuge the cell suspension at 200× g for 10 minutes, discard the supernatant, and resuspend the cells in 0.5 mL fresh StemPro34 medium (with cytokines) before adding them back to the plate.

3.4.1.3.4 Day 0: Perform Viral Transductions

1) Count the cells using the desired method (e.g., Countess automated cell counter), and calculate the volume of each virus needed to reach the target MOI using the *live* cell count and the titer information on the CoA (www.thermofisher.com/cytotune), following the equation in Table 3.1. *Note*: we recommend initially performing the transductions with MOIs of 5, 5, and 3 (i.e., KOS MOI = 5, hc-Myc MOI = 5, hKlf4 MOI = 3). These MOIs can be optimized for your application.

2) Harvest the cells and seed the necessary number of wells of a 24-well plate in a minimal volume (~100 μL) with 1.0×10^5 cells/well for each transduction.

3) Remove one set of CytoTune 2.0 Sendai tubes from the −80 °C storage. Thaw each tube one at a time by first immersing the bottom of the tube in a 37 °C water bath for 5–10 seconds, and then removing the tube from the water bath and allowing it to thaw at room temperature. Once thawed, briefly centrifuge the tube and place it immediately on ice.

4) Add the calculated volumes of each of the three CytoTune 2.0 Sendai viruses to 0.4 mL of prewarmed StemPro-34 medium containing cytokines and 4 μg/mL of Polybrene. Ensure that the solution is thoroughly mixed by pipetting the mixture gently up and down. Complete the next step within 5 minutes.

5) Add the reprogramming virus mixture to the well(s) containing cells. Incubate the cells at 37 °C in a humidified atmosphere of 5% CO_2 overnight.

3.4.1.3.5 Day 1: Replace Medium and Culture Cells

6) Remove the CytoTune 2.0 Sendai viruses by centrifuging the cells at 200× g for 10 minutes. Aspirate and discard the supernatant.
7) Resuspend the cells in 0.5 mL of complete StemPro-34 medium containing cytokines in the desired well of a 24-well plate.
8) Incubate the cells in at 37 °C in a humidified atmosphere of 5% CO_2 for 2 days.
9) Next, proceed with Option A *or* Option B.

3.4.1.3.6 Option A: Feeder-Dependent Reprogramming
Day 2: Prepare iMEF Culture Dishes

1) Prepare iMEF culture dishes as described previously. Typically six-well dishes are used for iPSC generation.

Day 3: Plate Transduced Cells on iMEF Culture Dishes

2) Count the cells using the desired method (e.g., Countess automated cell counter) and seed the six-well MEF culture plates with 10 000 to 50 000 *live* cells per well in 2 mL of complete StemPro-34 medium without the cytokines. Plate any excess cells in an additional iMEF culture dishes. *Note*: if there are leftover cells, they should be used as a positive control when performing RT-PCR or qPCR to determine presence of Sendai virus in the resulting iPSC. Freeze a cell pellet at −80 °C, or resuspend in 0.5 mL of TRIzol and store at −80 °C.
3) Incubate the cells at 37 °C in a humidified atmosphere of 5% CO_2.

Day 4 to Day 6: Replace Spent Medium

4) Every other day, gently remove 1 mL (half) of the spent medium from the cells and replace it with 1 mL of fresh complete StemPro-34 medium (without cytokines), without disturbing cells. *Note*: if cells are present in the 1 mL removed

from the wells, centrifuge the cell suspension at 200× g for 10 minutes, discard the supernatant, and resuspend the cells in 1 mL fresh StemPro34 medium (without cytokines) before adding them back to the plate.

Day 7: Start Transitioning Cells to PSC Medium

5) Remove 1 mL (half) of StemPro-34 medium from the cells and replace it with 1 mL of PSC medium to start the adaptation of the cells to the new culture medium.

Day 8 to Day 21: Feed and Monitor the Cells

6) Twenty-four hours later (day 8), change the full volume of the medium to PSC medium, and replace the spent PSC medium every day thereafter.
7) Starting on day 8, observe the plates every other day under a microscope for the emergence of cell clumps indicative of reprogrammed cells.
8) By day 15–21 after transduction, colonies should have grown to an appropriate size for transfer. The day before transferring the colonies, prepare iMEF culture in six-well dishes. *Note*: we typically harvest colonies closer to 21 days or less, to avoid differentiation.
9) When colonies are ready for transfer, perform live staining using Tra1-60 or Tra1-81 for selecting reprogrammed colonies if desired.
10) Manually pick colonies and transfer them onto freshly prepared iMEF plates.

3.4.1.3.7 *Option B: Feeder-Independent Reprogramming*
Day 3: Plate Transduced Cells

1) Prepare VTN or Geltrex coated culture dishes as described previously. Typically six-well dishes are used for iPSC generation.
2) Count the cells using the desired method (e.g., Countess automated cell counter) and seed the six-well VTN or Geltrex coated culture plates with 10 000 to 50 000 *live* cells per well in 2 mL of complete StemPro-34 medium (without cytokines). Plate any excess cells in an additional culture dishes.

Note: if there are leftover cells, they should be used as a positive control when performing RT-PCR or qPCR to determine presence of Sendai virus in the resulting iPSC. Freeze a cell pellet at −80 °C, or resuspend in 0.5 mL of TRIzol and store at −80 °C.

3) Incubate the cells at 37 °C in a humidified atmosphere of 5% CO_2.

Day 4 to Day 6: Replace Spent Medium

4) Every other day, gently remove 1 mL (half) of the spent medium from the cells and replace it with 1 mL of fresh complete StemPro-34 medium without cytokines and without disturbing cells. *Note*: if cells are present in the 1 mL removed from the wells, centrifuge the cell suspension at 200× g for 10 minutes, discard the supernatant, and resuspend the cells in 1 mL fresh StemPro34 medium (without cytokines) before adding them back to the plate.

Day 7: Start Transitioning Cells to Essential 8 Medium

5) Remove 1 mL (half) of StemPro-34 medium from the cells and replace it with 1 mL of Essential 8 medium to start the adaptation of the cells to the new culture medium.

Day 8–21: Feed and Monitor the Cells

6) Twenty-four hours later (day 8), change the full volume of the medium to Essential 8 medium, and replace the spent Essential 8 medium every day thereafter.

7) Starting on day 8, observe the plates every other day under a microscope for the emergence of cell clumps indicative of reprogrammed cells.

8) By day 15–21 after transduction, colonies should have grown to an appropriate size for transfer. The day of transferring the colonies, prepare VTN or Geltrex coated culture six-well plates. *Note*: we typically harvest colonies closer to 21 days or less, to avoid differentiation.

9) When colonies are ready for transfer, perform live staining using Tra1-60 or Tra1-81 for selecting reprogrammed colonies if desired.

10) Manually pick colonies and transfer them onto prepared culture plates coated with desired matrix.

3.4.1.4 Trouble Shooting: Optimization of Transduction with CytoTune-EmGFP Sendai Fluorescence Reporter

The CytoTune-EmGFP Sendai Fluorescence Reporter is a fluorescent control vector carrying the Emerald Green Fluorescent Protein (EmGFP) gene. The fluorescent control vector allows for determination of whether a cell line of interest is amenable or refractive to transduction by Sendai vectors. However, ability to be transduced by the reporter does not indicate the cell line's capability to be reprogrammed.

Please note the following considerations.

- You cannot transduce cells that have already been transduced with the CytoTune-EmGFP Sendai Fluorescence Reporter with CytoTune reprogramming vectors, and vice versa. If you wish to use the CytoTune-EmGFP Sendai Fluorescence Reporter during reprogramming, you must add it to the cells at the same time as the reprogramming vectors.
- Different cell types require different MOIs to express detectable levels of EmGFP. As such, cells should be transduced using a range of different MOIs. We suggest initially transducing your cells with at least 2–3 different MOIs (e.g., 1, 3, and 9).
- Expression of EmGFP should be detectable at 24 hours post transduction by fluorescence microscopy, and reach maximal levels at 48–72 hours.

3.4.1.4.1 Day –1 to –2: Prepare the Cells for Transduction

1) 1–2 days before transduction, plate the cells of interest onto the necessary number of wells of a multiwell plate at the appropriate density to achieve 50–80% confluency on the day of transduction (Day 0). One extra well can be used to count cells for viral volume calculations.
2) Culture the cells for 1–2 more days, ensuring the cells have fully adhered and extended.

3.4.1.4.2 Day 0: Perform Transduction

3) On the day of transduction, warm an appropriate volume of cell culture medium for each well to be transduced (e.g., 0.5 mL for each well of a 12-well plate) in a 37 °C water bath.
4) Harvest cells from one well of the multiwell plate and perform a cell count. These cells will not be transduced, but will

be used to estimate the cell number in the other well(s) plated in Step 1. *Note*: this step is optional and is performed to obtain more accurate MOI calculations. If exact MOIs are not needed, a rough estimate of the number of cells in the well (based on plating density and growth rates) will also suffice.

5) Count (or estimate) the cell number using the desired method (e.g., Countess automated cell counter), and calculate the volume of the virus needed to reach the target MOI(s), using the equation in Table 3.1. Titer information can be found on the CoA.

6) Remove one tube of CytoTune-EmGFP Sendai Fluorescence Reporter from the −80 °C storage. Thaw the vector by first immersing the bottom of the tube in a 37 °C water bath for 5–10 seconds, and then removing the tube from the water bath and allowing its contents to thaw at room temperature. Once thawed, briefly centrifuge the tube and place it immediately on ice.

7) Add the calculated volume of CytoTune-EmGFP Sendai Fluorescence Reporter to the prewarmed cell culture medium prepared in Step 3. Ensure that the solution is thoroughly mixed by pipetting the mixture gently up and down. Complete the next step within 5 minutes.

8) Aspirate the cell culture medium from the cells, and add the solution prepared in Step 7 to the well. Incubate the cells in a 37 °C, 5% CO_2 incubator overnight.

3.4.1.4.3   Day 1: Replace Medium and Culture Cells

9) Twenty-four hours after transduction, replace the medium with fresh cell culture medium. *Note*: depending on your cell type, you may expect to see some cytotoxicity 24–48 hours post transduction, which can affect >50% of your cells. This is an indication of high uptake of the virus. We recommend that you continue culturing your cells and proceed with the protocol.

10) Visualize the cells on a fluorescence microscope using a standard FITC filter set. EmGFP expression should be visible in some cells (expression will reach maximum levels at 48–72 hours).

3.4.1.4.4   Day 2+: Replace Medium and Culture Cells

11) Forty-eight hours after transduction, replace the medium with fresh cell culture medium.
12) Visualize the cells on a fluorescence microscope using a standard FITC filter set. EmGFP expression should be much brighter than Day 1, and should be visible in many cells (Figure 3.6).

3.4.1.5   Measuring Dilution of CytoTune Vectors

Dilution and presence of the Sendai vectors can be monitored via RT-PCR using the primers listed in Table 3.2. The SeV primer will detect the presence of any of the CytoTune Sendai vectors. The other primers will specifically detect the expression of the respective reprogramming transgenes, and will not cross-react with endogenous expression of these genes.

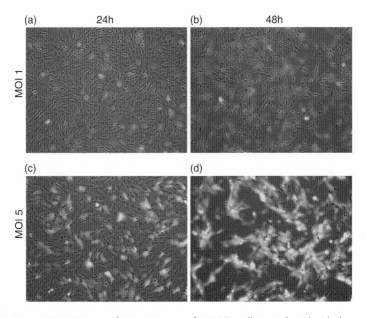

Figure 3.6 50× magnification images of BJ HDFn cells transduced with the CytoTune-EmGFP Sendai Fluorescence Reporter at the indicated MOI 1 (a,b) or MOI 5 (c,d) and imaged at the indicated time post-transduction – 24 h (a,c) or 48 hours (b,d).

Table 3.2 RT-PCR primers to detect CytoTune Sendai vectors.

Target	Primer set		Product size
SeV	Forward	GGA TCA CTA GGT GAT ATC GAG C	181 bp
	Reverse	ACC AGA CAA GAG TTT AAG AGA TAT GTA TC	
KOS	Forward	ATG CAC CGC TAC GAC GTG AGC GC	528 bp
	Reverse	ACC TTG ACA ATC CTG ATG TGG	
Klf4	Forward	TTC CTG CAT GCC AGA GGA GCC C	410 bp
	Reverse	AAT GTA TCG AAG GTG CTC AA	
c-Myc	Forward	TAA CTG ACT AGC AGG CTT GTC G	532 bp
	Reverse	TCC ACA TAC AGT CCT GGA TGA TGA TG	

3.4.1.5.1 *Total RNA Preparation*

1) Incubate the lysate with TRIzol Reagent at room temperature for 5 minutes to allow complete dissociation of nucleoprotein complexes.
2) To the TRIzol lysate add 0.2 mL chloroform per 1mL of TRIzol reagent and shake the tube vigorously for 15 seconds.
3) Incubate at room temperature for 2–3 minutes and centrifuge at $12\,000\times$ g for 15 minutes at 4 °C.
4) Carefully remove the upper aqueous phase and transfer to a new tube.
5) Add 0.5 mL 100% isopropanol to the aqueous phase per 1 mL of TRIzol reagent; incubate at room temperature for 10 minutes.
6) Centrifuge at $12\,000\times$ g for 10 minutes at 4 °C.
7) Carefully remove the supernatant from the RNA pellet and wash with 1 mL of 75% ethanol.
8) Centrifuge the tube at $7500\times$ g for 5 minutes at 4 °C. Discard the supernatant and air dry the RNA pellet for 5–10 minutes.
9) Resuspend the RNA pellet with 20–50 µL RNase-free water.

3.4.1.5.2 DNase Treatment

10) Add 0.1 times the RNA volume of 10× DNase I Buffer and 1 µL rDNase I to the RNA in a clean DNase/RNase-free micro-centrifuge tube, and mix gently.
For a 50 µL reaction:

RNA Sample	1–10 µg
10X DNase I Reaction Buffer	5 µL
rDNase I (2 Units)	1 µL
DEPC-treated water to bring reaction to 50 µL	X µL
Total	**50 µL**

11) Incubate the tube at 37 °C for 20–30 minutes.
12) Add the resuspended DNase Inactivation Reagent (typically 0.1 times the RNA volume) and mix well.
13) Incubate 2 minutes at room temperature, mixing occasionally.
14) Centrifuge at 10 000× g for 1.5 minutes and transfer the RNA to a fresh tube.

3.4.1.5.3 RNA Quantification

15) Use Nano Drop to quantify the extracted RNA sample. Quality of RNA is best assessed using Absorbance 260/280 ratio, with the recommended value close to 2.0.
16) RNA integrity can be further assessed by running the samples on a 1% agarose gel and assessing the 2:1 ratio of the 28s and 18s RNA bands and the absence of degraded RNA that appears as a small molecular weight smear.
17) If using a Bioanalyzer, a RIN (RNA integrity number) value of higher than 5 may be sufficient, but higher than 8 is ideal for downstream applications.

3.4.1.5.4 Reverse Transcription

18) For a single reaction, combine the following components in a tube on ice. For multiple reactions, prepare a master mix without RNA.

5× VILO Reaction Mix	4 µL
10× SuperScript Enzyme mix	2 µL
RNA (up to 2.5 µg)	X µL
DEPC-treated water to	20 µL

19) Gently mix tube contents and incubate at 25 °C for 10 minutes.
20) Incubate tube at 42 °C for 60 minutes.
21) Terminate the reaction at 85 °C for 5 minutes.
22) Use diluted or undiluted cDNA in PCR.

3.4.1.5.5 PCR

23) Carry out PCR using 10 µL cDNA from the RT reaction with the AccuPrime SuperMix according to Table 3.3.
24) Cap or seal the tube/plate and tap gently to mix and centrifuge briefly to collect the contents.
25) Place the tubes/plates in the thermal cycler and run the program on the StepOne q-RT-PCR machine using the parameters in Table 3.4.
26) Analyze the PCR products on a 2% agarose gel.

Table 3.3 Volumes for PCR reaction (Sendai vector detection).

Component	10 µL reaction	25 µL reaction	50 µL reaction
Accuprime SuperMix I	5 µL	12.5 µL	25 µL
Primer Mix (10 µM each)	0.2 µL	0.5 µL	1 µL
Template DNA	1–200 ng	1–200 ng	1–200 ng
DNase-free water	Up to 10 µL	Up to 25 µL	Up to 50 µL

Table 3.4 Conditions for PCR reaction (Sendai vector detection).

Step	Temperature	Time	Cycles
Denaturing	95 °C	30 seconds	30–35 cycles
Annealing	55 °C	30 seconds	
Template DNA	72 °C	30 seconds	

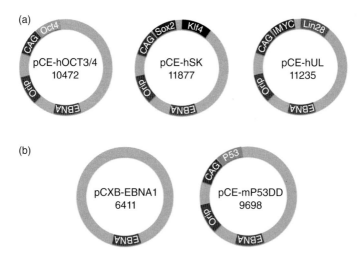

Figure 3.7 Epi5 reprogramming kit comprising five vectors in two tubes. Reprogramming vector tube containing a mixture of three plasmids that code for OCT3/4, SOX2, KLF4, L-MYC, and LIN28 (a); second tube containing a mixture of two plasmids that code for the p53 dominant negative mutant and EBNA1 (b).

3.4.2 Episomal Reprogramming

The Epi5 Episomal iPSC Reprogramming Kit provides an easy-to-use, highly efficient set of five episomal vectors that deliver the necessary factors for integration free reprogramming of human somatic cells. This system produces transgene-free, virus-free human iPSCs without the need for small molecules during reprogramming and the added p53 suppression provides enhanced iPSC generation (Figure 3.7). As oriP/EBNA1 vectors, these episomal vectors contain five reprogramming factors (Oct4, Sox2, Lin28, Klf4, and L-Myc) and replicate extrachromosomally only once per cell cycle. The following protocols describe the reprogramming of human fibroblasts and CD34+ blood cells using Epi5 vectors.

3.4.2.1 Reprogramming Human Fibroblast with Epi5
3.4.2.1.1 *Day −3: Seed Cells*

1) Three days before transduction, remove one vial of BJ fibroblasts from the liquid nitrogen storage tank.
2) Briefly roll the cryovial between hands to remove frost, and swirl it gently in a 37 °C water bath to thaw the cells.

3) When only a small ice crystal remains in the vial, remove it from water bath. Spray the outside of the vial with 70% ethanol before placing it in the cell culture hood.

4) Pipette the thawed cells gently into a 50 mL conical tube.

5) Add 10 mL of prewarmed fibroblast medium drop-wise to the cells. Gently mix by pipetting up and down. *Note*: adding the medium slowly helps the cells to avoid osmotic shock.

6) Transfer the contents to a 15 mL conical tube and centrifuge the cell suspension at 200× g for 4 minutes.

7) Gently discard the supernatant and resuspend the cells in 1 mL of complete fibroblast medium.

8) Typically seed the contents of the vial in 15 mL of medium in a T-75 TC treated culture flask.

9) Incubate the BJ fibroblast culture in a 37 °C, 5% CO_2 incubator overnight.

10) The following day, aspirate off the spent medium and replace with fresh, prewarmed fibroblast medium.

11) Change the medium every other day until the flask is approximately 70% confluent (usually 4–5 days).

3.4.2.1.2 Day 0: Transfect Cells Using the Neon Transfection System

Note: Gentle handling of the cells prior to transfection is essential for the success of the transfection procedure.

12) Wash the culture flask twice with DPBS (-/-).

13) Aspirate off the D-PBS wash and add appropriate volume of TrypLE Select reagent or 0.05% trypsin/EDTA. Incubate the cells for 2–5 minutes at 37 °C.

14) When the cells have rounded up, add fibroblast medium into each well to stop the trypsinization.

15) Dislodge the cells from the well by gently pipetting the media across the surface of the dish. Collect the cell suspension in a 15 mL conical centrifuge tube.

16) Centrifuge the cells at 200× g for 2 minutes.

17) Aspirate off the supernatant, and resuspend the cells in an appropriate amount of fresh, prewarmed fibroblast medium.

18) Count the cells using the desired method (e.g., Countess automated cell counter or hemacytometer), and transfer cells to microfuge tube at a density of $1-5 \times 10^5$ cells and centrifuge at 200× g for 4 minutes.

19) Carefully aspirate most of the supernatant, leaving approximately 100–200 µL behind. Remove the remaining medium with a 200 µL pipette.

20) Resuspend the cell pellet in Re-suspension Buffer R (included with Neon Transfection kits) at a final concentration of $1.0 \times 10^7 - 1.4 \times 10^7$ cells/ mL.

21) Transfer the cells (10 µL per transfection reaction) to a sterile 1.5 mL microcentrifuge tube.

22) Fill the Neon Tube with 3 mL electrolytic buffer (use Buffer E for the 10 µL Neon Tip).

23) Insert the Neon Tube into the Neon Pipette Station until you hear a click.

24) Transfer 1 µL of Epi5 Reprogramming Vectors to the tube containing cells.

25) Transfer 1 µL of Epi5 p53 and EBNA Vectors to the tube containing cells and mix gently.

26) Insert a 10 µL Neon Tip into the Neon Pipette.

27) Press the push-button on the Neon Pipette to the first stop and immerse the Neon Tip into the cell-DNA mixture. Slowly release the push-button on the pipette to aspirate the cell-DNA mixture into the Neon Tip. *Note*: avoid air bubbles during pipetting to avoid arcing during electroporation. If you notice air bubbles in the tip, return the sample to the microcentrifuge tube, and carefully draw the sample into the tip again without any air bubbles.

28) Insert the Neon Pipette with the sample vertically into the Neon Tube placed in the Neon Pipette Station until you hear a click.

29) Ensure that you have entered the appropriate electroporation parameters (Table 3.5) and press Start on the Neon touchscreen to deliver the electric pulse. *Note*: during electroporation, watch the Neon Tip carefully for arcing, which appears as a brief, bright flash or spark. If arcing is observed,

Table 3.5 Neon settings for Epi5 electroporation of human dermal fibroblasts or CD34+ cells.

Pulse voltage	Pulse number	Pulse width	Cell density	Tip size
1650 V	3	10 ms	1×10^7 cells/mL	10 µL

discard the sample and start again at Step 21. After the electric pulse is delivered, the touchscreen displays "Complete" to indicate that electroporation is complete.

30) Remove the Neon Pipette from the Neon Pipette Station and immediately transfer the samples from the Neon Tip into one well of the prewarmed Geltrex or vitronectin matrix coated six-well plate containing 2 mL of fibroblast media. *Note*: evenly distribute cells over the well in a drop-wise manner.

31) Discard the Neon Tip into an appropriate biological hazardous waste container.

32) Repeat the process for any additional samples. Do not use a Neon Tip more than twice. If the same cells are being used for each reaction, the Neon Tube and Buffer 4 may be used up to 10 times.

33) Incubate the plates at 37 °C in a humidified CO_2 incubator overnight.

3.4.2.1.3 *Day 1: Switch to N2/B27 Medium*

34) Twenty-four hours after electroporation, carefully aspirate the fibroblast medium from the top of the well.

35) Add 1 mL of N2/B27 medium supplemented with 100 ng/mL of bFGF, to each well.

3.4.2.1.4 *Day 2 to Day 13: Replace Spent N2/B27 Medium*

36) Each day, carefully aspirate the spent medium and replace with 2 mL of N2/B27 medium supplemented with 100 ng/mL bFGF.

37) Replace spent medium with 2 mL of fresh N2/B27 medium, supplemented with 100 ng/mL of bFGF, every day, up to day 14 post transfection.

3.4.2.1.5 *Day 14 to Day 21: Switch to Essential 8 Medium*

38) Aspirate off the spent N2/B27 medium and replace it with complete Essential 8 medium. Change Essential 8 medium every day.

39) Observe the plates every other day under a microscope for the emergence of cell clumps indicative of transformed cells. Within 15–21 days of transfection, the iPSC colonies will grow to an appropriate size for transfer.

3.4.2.2 Reprogramming CD34+ Cells with Epi5
3.4.2.2.1 Day –3 to –1: Seed and Expand Cells

1) Three days before transduction, remove one vial of CD34+ from the liquid nitrogen storage tank.
2) Briefly roll the cryovial between hands to remove frost, and swirl it gently in a 37 °C water bath to thaw the CD34+ cells.
3) When only a small ice crystal remains in the vial, remove it from water bath. Spray the outside of the vial with 70% ethanol before placing it in the cell culture hood.
4) Pipette the thawed cells gently into a 15 mL conical tube.
5) Add 10 mL of prewarmed complete StemPro-34 medium drop-wise to the cells. Gently mix by pipetting up and down. *Note*: adding the medium slowly helps the cells to avoid osmotic shock.
6) Centrifuge the cell suspension at 200× g for 10 minutes.
7) Gently discard the supernatant and resuspend the cells to a density of 5×10^5 cells/mL, using the appropriate amount of StemPro-34 medium containing cytokines (i.e., SCF, IL3, and GM-CSF).
8) Place 0.5 mL of cell suspension into each well of a 24-well plate and incubate at 37 °C in a humidified atmosphere of 5% CO_2. *Note*: we recommend using the wells in the middle section of the 24-well plate to prevent excessive evaporation of the medium during incubation.
9) Two days before transduction (day –2), add 0.5 mL of fresh complete StemPro-34 medium (containing cytokines) without disturbing the cells. *Note*: if cells are present in the 0.5 mL removed from the wells, centrifuge the cell suspension at 200× g for 10 minutes, discard the supernatant, and resuspend the cells in 0.5 mL fresh StemPro34 medium (with cytokines) before adding them back to the plate.
10) One day before transduction (day –1), gently remove 0.5 mL of the medium, and add 1.0 mL of fresh complete StemPro-34 medium (containing cytokines) without disturbing the cells.

3.4.2.2.2 Day 0: Transfect Cells Using the Neon Transfection System

11) Harvest all the cells from the well, use additional medium to harvest any remaining cells. *Note*: gentle handling of the cells prior to transfection is essential for the success of the transfection procedure.

12) Centrifuge the cell suspension at 200× g for 10 minutes, discard the supernatant, and resuspend the cells in 1 mL of complete StemPro-34 medium containing cytokines (i.e., SCF, IL3, and GM-CSF)

13) Count the number of cells using a hemacytometer or the Countess automated cell counter to determine the viability and total number of cells.

14) Carefully aspirate most of the supernatant, leaving approximately 100–200 μL behind. Remove the remaining medium with a 200 μL pipette.

15) Resuspend the cell pellet in Resuspension Buffer T (included with Neon Transfection kits) at a final concentration of 1.0×10^7 –1.4×10^7 cells/ mL. *Note*: Buffer T is required for blood cell transfections.

16) Transfer the cells (10 μL per transfection reaction) to a sterile 1.5 mL microcentrifuge tube.

17) Fill the Neon Tube with 3 mL Electrolytic Buffer (use Buffer E for the 10 μL Neon Tip).

18) Insert the Neon Tube into the Neon Pipette Station until you hear a click.

19) Transfer 1 μL of Epi5 Reprogramming Vectors to the tube containing cells.

20) Transfer 1 μL of Epi5 p53 and EBNA Vectors to the tube containing cells and mix gently.

21) Insert a 10 μL Neon Tip into the Neon Pipette.

22) Press the push-button on the Neon Pipette to the first stop and immerse the Neon Tip into the cell-DNA mixture. Slowly release the push-button on the pipette to aspirate the cell-DNA mixture into the Neon Tip. *Note*: avoid air bubbles during pipetting to avoid arcing during electroporation. If you notice air bubbles in the tip, return the sample to the microcentrifuge tube, and carefully draw the sample into the tip again without any air bubbles.

23) Insert the Neon Pipette with the sample vertically into the Neon Tube placed in the Neon Pipette Station until you hear a click.

24) Ensure that you have entered the appropriate electroporation parameters (see Table 3.5) and press Start on the Neon touchscreen to deliver the electric pulse. *Note*: during electroporation, watch the Neon Tip carefully for arcing, which

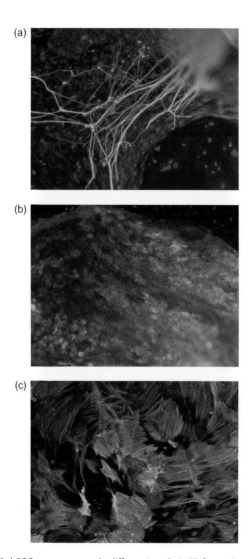

(a)

(b)

(c)

Figure 4.8 hPSCs spontaneously differentiated via EB formation and allowed to mature in culture for 21 days in order to characterize trilineage potential as a measure of pluripotency. Cultures are then probed with antibodies via ICC for cells indicative of the three somatic germ lineages: the neuronal marker beta III tubulin for ectoderm (a), the hepatic marker alpha-fetoprotein (AFP) for endoderm (b), and the cardiac marker smooth muscle actin (SMA) for mesoderm (c).

Human Pluripotent Stem Cells: A Practical Guide, First Edition. Uma Lakshmipathy, Chad C. MacArthur, Mahalakshmi Sridharan and Rene H. Quintanilla. © 2018 John Wiley & Sons, Inc. Published 2018 by John Wiley & Sons, Inc.

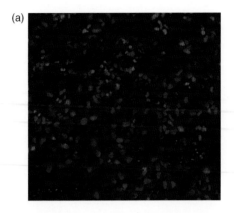

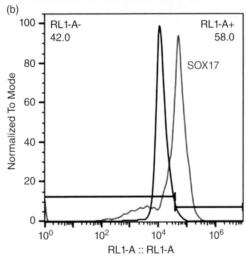

Figure 5.2 Definitive endoderm cells derived from iPSCs hPSCs induced to definitive endoderm, at Day 3. ICC performed using FOXA2 (*red*), counterstained for nuclei (*blue*) with DAPI (a). Flow analysis for Sox17 staining of PSC (*black*) and PSC-derived definitive endoderm (*red*) (b).

(a)

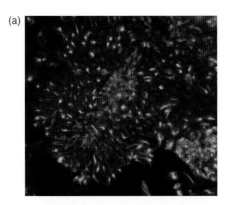

(b)

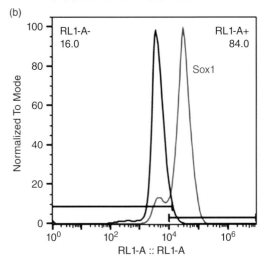

Figure 5.4 Neural stem cells derived from PSCs. hPSCs induced to NSCs. ICC performed using Nestin (*green*), Sox2 (*red*), counterstained for nuclei (*blue*) with DAPI (a). Flow analysis of Sox1 staining of PSCs (*black*) and PSC-derived NSCs (*red*) (b).

(a)

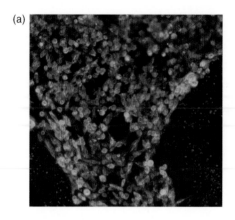

(b)

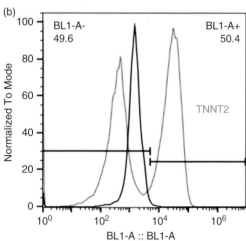

Figure 5.6 Cardiomyocytes derived from PSCs. hPSCs induced to cardiomycotyes. ICC performed using TNNT2 (*green*) and NKX2.5 (*red*) counterstained for nuclei (*blue*) with DAPI (a). Flow analysis of TNNT2 staining of PSCs (*black*) and PSC-derived cardiomyocytes (*green*) (b).

appears as a brief, bright flash or spark. If arcing is observed, discard the sample and start again at Step 16. After the electric pulse is delivered, the touchscreen displays "Complete" to indicate that electroporation is complete.

25) Remove the Neon Pipette from the Neon Pipette Station and immediately transfer the samples from the Neon Tip into one well of the prewarmed Geltrex matrix coated six-well plate containing 2 mL of complete StemPro-34 medium with cytokines. *Note*: evenly distribute cells over the well in a drop-wise manner.

26) Discard the Neon Tip into an appropriate biological hazardous waste container.

27) Repeat the process for any additional samples. Do not use a Neon Tip more than twice. If the same cells are being used for each reaction, the Neon Tube and Buffer E may be used up to 10 times.

28) Incubate the plates at 37 °C in a humidified CO_2 incubator overnight.

3.4.2.2.3 Day 1: Switch to N2/B27 Medium

29) Twenty-four hours after electroporation, carefully remove 0.5 mL of the supernatant from the top of the well, trying to not remove any of the suspension CD34+ cells while they are still attaching.

30) Add 1 mL of N2/B27 medium supplemented with 100 ng/mL of bFGF, to each well.

3.4.2.2.4 Day 2 to Day 8: Replace Spent N2/B27 Medium

31) Each day, carefully remove the spent medium with a 5 mL pipette, and replace with 2 mL of N2/B27 medium supplemented with 100 ng/mL bFGF.

32) Replace spent medium with 2 mL of fresh N2/B27 medium, supplemented with 100 ng/mL bFGF, every day, up to day 9 post transfection.

3.4.2.2.5 Day 9 to Day 21: Switch to Essential 8 Medium

33) Aspirate off the spent N2/B27 medium and replace it with complete Essential 8 medium. Change Essential 8 medium every other day.

34) Observe the plates every other day under a microscope for the emergence of cell clumps indicative of transformed cells. Within 15–21 days of transfection, the iPSC colonies will grow to an appropriate size for transfer.

3.4.2.3 PCR to Detect Epi5 Vectors

The presence of Epi5 Vectors in reprogrammed iPSC colonies can be detected by endpoint PCR, using the PCR primers listed in Table 3.6. The EBNA-1 primer set can detect all five episomal plasmids in the kit. The oriP primer set can detect all episomal plasmids in the kit except for pCXBEBNA1, which lacks the OriP gene.

3.4.2.3.1 *Harvesting iPSC for DNA*

1) Harvest iPSC using standard methods as described in Chapter 2 (e.g., collagenase for feeder-dependent culture or EDTA for feeder-free culture).
2) Once cells are collected and centrifuged, aspirate and discard the supernatant. Resuspend the cell pellet in 500 μL DPBS and transfer resuspended cells to a thin-walled 0.5 mL PCR tube.
3) Centrifuge the cell suspension at 200× g for 5 minutes to pellet cells.
4) Aspirate and discard the supernatant. Resuspend the cell pellet in 20 μL of Resuspension Buffer with 2 μL of Lysis Solution added to the Resuspension Buffer.

Table 3.6 PCR primers for detection of Epi5 vectors.

Transgene	Primers	Sequence	Expected size
oriP	pEP4-SF1-oriP	5′-TTC CAC GAG GGT AGT GAA CC-3′	544 bp
	pEP4-SR1-oriP	5′-TCG GGG GTG TTA GAG ACA AC-3′	
EBNA-1	pEP4-SF2-oriP	5′-ATC GTC AAA GCT GCA CAC AG-3′	666 bp
	pEP4-SR2-oriP	5′-CCC AGG AGT CCC AGT AGT CA-3′	

5) Incubate the cells for 10 minutes in an incubator or thermal cycler that has been preheated to 75 °C.
6) Centrifuge the tube briefly to collect any condensation. Use 3 μL of the cell lysate in a 50 μL PCR reaction.

3.4.2.3.2   PCR Reaction

1) Add the components listed in Table 3.7 to a DNase/RNase-free, thin-walled PCR tube. For multiple reactions, prepare a master mix of common components to minimize reagent loss and enable accurate pipetting. *Note*: assemble PCR reactions in a DNA-free environment. We recommend use of clean dedicated automatic pipettors and aerosol resistant barrier tips.
2) Cap the tube, tap gently to mix, and centrifuge briefly to collect the contents.
3) Place the tube in the thermal cycler and use the PCR parameters shown in Table 3.8.
4) Analyze the PCR products using 2% agarose gel electrophoresis.

3.4.3   Identify and Isolate iPSC Colonies

1) Three to four weeks after transduction, colonies should have grown to an appropriate size for transfer. The day before transferring the colonies, prepare iMEF culture plates for

Table 3.7 Volumes for PCR (Epi5 vector detection).

Component	Volume per reaction (μL)
10× PCR Buffer II	5
Forward Primer (10 μM stock)	1
Reverse Primer (10 μM stock)	1
Accuprime Taq Polymerase (5 units/μL)	1
Cell lysate	3
Sterile distilled water	39

Table 3.8 Conditions for PCR (Epi5 vector detection).

Step	Temperature	Time	Cycles
Initial denaturation	94 °C	2 minutes	–
Denaturation	94 °C	30 seconds	35–40
Annealing	55 °C	30 seconds	
Elongation	72 °C	1 minute	
Final elongation	72 °C	7 minutes	–

feeder-dependent reprogramming. For feeder-independent reprogramming, VTN or Geltrex coated dishes can be prepared the day of transfer. The size of the plates to be used to pick colonies will depend on your particular need; we recommend using six-well plates.

2) Manually score and pick individual colonies and transfer them onto the appropriate prepared plates, one scored colony per well to ensure clonal expansion.

3) In addition to using morphological cues to identifying iPSC colonies, we recommend using Alkaline Phosphatase Live Stain or Tra-1-60 Kit for live cell imaging (Figure 3.8).

4) Once desired iPSC colonies have been selected and transferred, replace the iPSC medium (feeder-dependent) or Essential 8 medium (feeder-independent) daily, and culture as any other feeder-dependent PSC, using the techniques described in previous sections. *Note*: early passage iPSCs may not survive passage by traditional enzymatic methods, and should be manually passaged for the first few passages. It is very important to expand your clonal iPSCs and characterize them, as some picked colonies may be partially reprogrammed or may not survive the initial picking.

3.4.4   Determining Reprogramming Efficiency

Terminal AP staining is also a valuable tool in determining the efficiency of a reprogramming process. The following protocol describes a terminal staining using the Vector Red Staining Kit from Vector Labs. It permits the accumulation of the substrate

(a) (b)

(c)

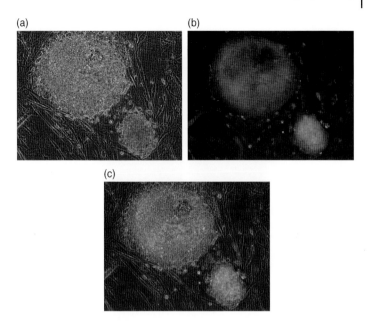

Figure 3.8 100× magnification images of feeder-dependent emerging iPSC colonies stained with AP Live Stain. Phase contrast (a), AP Live Stain (b), and merge of images (a) and (b) (c).

in the cells, which yields a red colorimetric signal, which is visible to the eye, as well as a fluorometric signal that can be detected in the TRIT-C/Cy3 channel. This method is non-reversible and destructive but is a good indicator of the efficiency of reprogramming somatic cells into iPSCs, as the number of AP positive colonies can be counted.

1) This protocol describes AP staining of reprogramming master plates in a six-well format, scale up as necessary. Prepare a 5 mL staining mix by adding the following reagents from the Vector Red kit, at room temperature. *Note*: once the staining mix is prepared, it is best to use it within 5–10 minutes to prevent degradation of the substrate.
   a) Add 5 mL of 200 mM Tris-HCL solution to a 15 mL conical tube.
   b) Add 2 drops of Reagent 1 per 5 mL of Tris solution and mix.
   c) Add 2 drops of Reagent 2 per 5 mL of Tris solution and mix.
   d) Add 2 drops of Reagent 3 per 5 mL of Tris solution and mix.

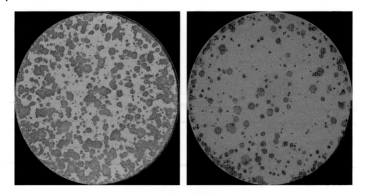

Figure 3.9 Terminal AP staining of BJ fibroblasts reprogrammed (day 21) with CytoTune 2.0, using feeders and KSR-based medium (*left*) and vitronectin and Essential 8 medium (*right*). Colonies can be manually counted.

2) Aspirate off the normal growth media from the wells of a six-well plate. *Note*: for determining reprogramming efficiencies, we recommend that the master dish be assessed at Day 21, or later, of the reprogramming process to ensure the colonies are large enough to properly measure.

3) Add 2 mL of the 200 mM Tris-HCL solution to each well and incubate for 2–3 minutes to equilibrate the samples before staining.

4) Aspirate off the Tris wash and add 2 mL of the diluted Vector Red AP stain and incubate for 20–30 minutes, at room temperature protected from light.

5) Following the incubation, aspirate off the staining solution and add 2 mL of the 200 mM Tris-HCL solution to each well and incubate for 3–5 minutes to remove any residual stain that may cause background signal.

6) Aspirate off the Tris wash and visualize the AP stain, either via normal photography or under red fluorescent analysis (Figure 3.9). *Note*: it is best to not leave any liquid in the wells, as the colonies may sometimes peel off. Colony counts can be performed by hand or via an automated counting device, such as the Incucyte Zoom imaging system.

References

1   K. Takahashi *et al.* Induction of pluripotent stem cells from adult human fibroblasts by defined factors. *Cell* **131**, 861–872 (2007).

2   J. Yu *et al.* Induced pluripotent stem cell lines derived from human somatic cells. *Science* **318**, 1917–1920 (2007).

3   N. Fusaki, H. Ban, A. Nishiyama, K. Saeki, M. Hasegawa. Efficient induction of transgene-free human pluripotent stem cells using a vector based on Sendai virus, an RNA virus that does not integrate into the host genome. *Proc Jpn Acad Ser B Phys Biol Sci* **85**, 348–362 (2009).

4   N. Yoshioka *et al.* Efficient generation of human iPSCs by a synthetic self-replicative RNA. *Cell Stem Cell* **13**, 246–254 (2013).

5   J. Yu *et al.* Human induced pluripotent stem cells free of vector and transgene sequences. *Science* **324**, 797–801 (2009).

6   K. Hu *et al.* Efficient generation of transgene-free induced pluripotent stem cells from normal and neoplastic bone marrow and cord blood mononuclear cells. *Blood* **117**, e109–119 (2011).

7   K. Okita *et al.* An efficient nonviral method to generate integration-free human-induced pluripotent stem cells from cord blood and peripheral blood cells. *Stem Cells* **31**, 458–466 (2013).

8   L. Warren *et al.* Highly efficient reprogramming to pluripotency and directed differentiation of human cells with synthetic modified mRNA. *Cell Stem Cell* **7**, 618–630 (2010).

9   T.M. Schlaeger *et al.* A comparison of non-integrating reprogramming methods. *Nat Biotechnol* **33**, 58–63 (2015).

4

Characterization

4.1   Introduction

Pluripotent stem cell (PSC) cultures are often derived using varying culture conditions from diverse genetic backgrounds that could have an impact on quality as measured by gene signatures and differentiation potential, and genome stability [1–5]. It is therefore good practice to carry out thorough characterization of cell banks with periodic characterization of PSCs in culture [6,7]. A typical characterization scheme of iPS starting from their generation to application is summarized in Figure 4.1.

The most basic method used to assess quality of PSCs is based on morphology but since this requires a trained eye, methods that allow live staining such as detection of surface markers with antibodies or use of live dyes for alkaline phosphatase can be used [8]. Undifferentiated PSCs are confirmed for expression of self-renewal markers using both cellular and molecular methods. Cellular methods include detection of surface markers such as SSEA4 and TRA-1-60 and intracellular markers such as Oct4 using immunostaining methods. Flow cytometry of these positive surface markers like TRA-1-60, in addition to absence of differentiation markers such as SSEA1, can provide an accurate measure of PSC quality. Immunostaining can be either a direct, single-step method using fluorescent-dye labeled antibodies or an indirect, two-step method requiring a primary antibody and a fluorophore-conjugated secondary antibody. Molecular methods are now being used to characterize PSCs through

Human Pluripotent Stem Cells: A Practical Guide, First Edition. Uma Lakshmipathy, Chad C. MacArthur, Mahalakshmi Sridharan and Rene H. Quintanilla.
© 2018 John Wiley & Sons, Inc. Published 2018 by John Wiley & Sons, Inc.

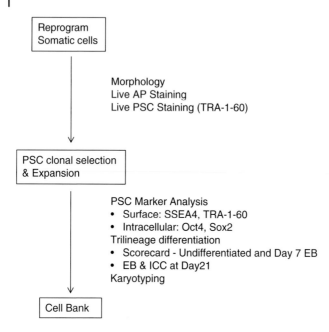

Figure 4.1 Schematic of comprehensive characterization.

transcriptome and epigenome analysis [1,9,10]. These analyses have led to identification of novel markers that can be used to distinguish fully reprogrammed iPSCs from partially reprogrammed and unreprogrammed cells [11]. A combination of positive and negative markers also enables tracking the progression of reprogramming [12].

In addition to confirmation of expression signature, PSCs have to be confirmed to possess trilineage differentiation potential. This is achieved by induction of spontaneous differentiation via embryoid body (EB) formation. After 21 days, these EBs are fixed and stained with antibodies specific for cell types representing the three germ layers – endoderm, mesoderm, and ectoderm. Alternatively, molecular analysis tools such as TaqMan hPSC ScoreCard enables confirmation of trilineage differentiation based on gene expression pattern at as early as 7 days of differentiation via analysis of a focused set of genes using PCR [13].

4.1.1   Additional Considerations

• Validate all antibodies used for specificity to the intended target via specific phenotypical marker expression patterns.
• If multiplexing antibodies, insure they do not cross-react with each other and are raised in different animals (i.e., mouse IgG marker 1 and rabbit IgG marker 2, with secondary antibodies against each animal using different fluorophores).

4.2   Materials

4.2.1   Embryoid Body Formation

1) DMEM/F-12 (1×), Liquid (1:1), with GlutaMAX™-I *Cat# 10565-018*
2) MEM Non-Essential Amino Acids Solution (100×) *Cat# 11140-050*
3) KnockOut™ Serum Replacement *Cat# 10828-010*
4) 2-Mercaptoethanol *Cat# 21985-023*
5) FGF-basic (AA 1-155) Recombinant Human *Cat# PHG0264*
6) Collagenase Type IV *Cat# 17104-019*
7) Dispase II, powder *Cat# 17105-041*
8) StemPro™ EZPassage™ Disposable Stem Cell Passaging Tool *Cat# 23181-010*
9) BD Falcon Cell Scraper Fisher *Cat# 08-771-1A*

4.2.2   Characterization

1) DMEM/F-12 (1X), Liquid (1:1), with GlutaMAX™-I *Cat# 10565-018*
2) SSEA4, Mouse Monoclonal Unconjugated Antibody *Cat# 414000*
3) Tra1-60, Mouse Monoclonal Unconjugated Antibody *Cat# 411000*
4) Tra1-81, Mouse Monoclonal Unconjugated antibody *Cat# 411100*
5) AlexaFluor™ 488 Goat Anti-mouse IGg (H+L) secondary Antibody *Cat# A11029*
6) AlexaFluor™ 594 Goat Anti-mouse IGg (H+L) secondary Antibody *Cat# A11020*

7) Alkaline Phosphatase Live Stain (500×) *Cat# A14353*
8) Dulbecco's Phosphate Buffered Saline (DPBS) without calcium and magnesium *Cat# 14190-144*
9) Paraformaldehyde, 4% Affymetrix *Cat# 19943*
10) Triton X-100 (10%) *Cat# 85111*
11) Goat serum *Cat# PCN5000*
12) Blocker™ BSA (10×) in PBS *Cat# 37525*
13) Rabbit anti-Oct4 antibody *Cat# A13998*
14) AlexaFluor™ 594 goat anti-rabbit IgG *Cat# A11037*
15) NucBlue™ Fixed Cell ReadyProbes™ Reagent *Cat# R37606*
16) ProLong™ gold anti-fade reagent *Cat# P36930*
17) Pluripotent Stem Cell 4-Marker Immunocytochemistry Kit *Cat# A24881*
18) Mouse Monoclonal primary anti-Human beta-3-Tubulin (b-3-Tub) *Cat# 480011*
19) Mouse Monoclonal primary anti-alpha-fetoprotein (AFP) *Cat# 180003*
20) Mouse Monoclonal primary anti-Smooth muscle Actin (SMA) *Cat# 180106*
21) 3-Germ Layer Immunocytochemistry Kit *Cat# A25538*

4.2.3   hPSC Flow Cytometry

1) Dulbecco's Phosphate Buffered Saline (DPBS) without calcium and magnesium *Cat# 14190-144*
2) TrypLE™ Express Enzyme *Cat# 12604-013*
3) BD Falcon Tube with Cell Strainer Cap Fisher Scientific *Cat# 08-771-23*
4) AlexaFluor 647 Mouse Anti-SSEA-1 BD Biosciences 100 tests 5 mL *Cat# 560120*
5) AlexaFluor 488 TRA-1-60 BD Biosciences 100 tests 5 mL *Cat# 560173*

4.2.4   TaqMan hPSC Scorecard from Total RNA

1) TRIzol™ *Cat# 5596-026*
2) Chloroform Sigma *Cat# C-2432*
3) Isopropanol Sigma *Cat# I9516-500ml*
4) Ethanol Sigma *Cat# E7023-500ml*
5) RNase-free Water *Cat# 10977*

6) DNA-fre Kit *Cat# AM1906*
7) High-capacity cDNA Reverse Transcription kit with RNase Inhibitor, 200 reactions *Cat# 4374966*
8) TaqMan® hPSC Scorecard™ Panel, 384-well *Cat# A15870*
9) TaqMan® hPSC Scorecard™ Kit, 384-well *Cat# A15872*
10) TaqMan® hPSC Scorecard™ Panel, Fast 96-well *Cat# A15876*
11) TaqMan® hPSC Scorecard™ Kit, Fast 96-well *Cat# A15871*
12) Micro Amp™ Optical Adhesive Film *Cat# 4311971*
13) TaqMan® Gene Expression Master Mix *Cat# 4369016*
14) TaqMan® Fast Advanced Master Mix *Cat# 4444558*

4.3 Solutions

4.3.1 Human PSC Medium (for 500 mL)

DMEM-F12	395 mL
KnockOut Serum Replacement	100 mL
NEAA	5 mL
2-Mercaptoethanol	500 L

Sterilize through 0.22 μm filter. Medium lasts for up to 28 days at 4 °C. Add bFGF (final concentration 4 ng/mL) fresh daily prior to use (example: 0.4 μL reconstituted bFGF per mL of medium).

4.3.2 Basic FGF Solution (10 μg/mL, for 1 mL)

Basic FGF	10 μg
D-PBS (-/-)	990 μL
KnockOut Serum Replacement	10 μL

Aliquot and store at −20 °C for up to 3 months. Once bFGF aliquot is thawed, use within 7 days, when stored at 4 °C.

4.3.3 Collagenase IV Solution (1 mg/mL, for 50 mL)

Collagenase IV	50 mg
DMEM-F12	50 mL

Sterilize through 0.22 μm filter and store at 4 °C for up to 14 days.

4.3.4 Dispase Solution (2 mg/mL, for 50 mL)

Dispase	100 mg
DMEM-F12	50 mL

Sterilize through 0.22 μm filter and store at 4 °C for up to 14 days.

4.3.5 Embryoid Body Medium (without bFGF) (for 500 mL)

DMEM-F12	395 mL
KnockOut Serum Replacement	100 mL
NEAA	5 mL
2-Mercaptoethanol	500 μL

Sterilize through a 0.22 μm filter unit. Medium lasts for up to 1 month at 4 °C.

4.3.6 ICC Blocking Buffer (for 100 mL)

D-PBS without $CaCl_2$ and $MgCl_2$	84 mL
Normal Goat serum	5 mL
10% Triton X-100	1 mL
10% BSA solution	10 mL

Sterilize through a 0.22 μm filter unit. Medium lasts for up to 1 month at 4 °C.

4.4 Methods

4.4.1 Alkaline Phosphatase Live Staining of hPSCs

Alkaline phosphatase (AP) is an enzyme that is upregulated in PSCs and can be detected using a substrate that selectively fluoresces as a result of AP activity. This method for the differential staining for AP activity is quick, reversible, and preserves the viability of the cells. Thus, it can be used to discriminate stem cells from feeder cells or parental cells during reprogramming.

1) Remove the normal PSC growth media from the desired sample well of a 12-well plate for routine characterizations (scale up for usage for iPSC colony selection from the master reprogramming plate or dish for clonal isolation of putative cell lines).

2) Gently add 500 μL of DMEM/F-12 medium, prewarmed to 37 °C, to the well and allow the wash solution to be incubated for 3 minutes at room temperature.

3) Aspirate off the DMEM/F-12 wash from the well and repeat once more. *Note*: it is crucial at this point to remove all traces of complete media that may contain FBS, KSR or BSA, which interferes with the reaction and signal acquisition.

4) Dilute the 500× AP Live Stain in prewarmed DMEM/F-12 to a final dilution of 1:500 (1 μL of substrate in 500 μL of DMEM/F-12) in a microcentrifuge tube and mix by gently pipetting up and down.

5) Aspirate off the final DMEM/F-12 wash and add the diluted AP Live Stain solution to the wells to be stained.

6) Incubate the cultures with the 1× AP Live Stain at 37 °C for 20–30 minutes in the 5% CO_2 humidified incubator.

7) Following the incubation period, aspirate off the AP Live Stain solution.

8) Add 500 μL of DMEM/F-12 to the well and allow the wash solution to be incubated for 3 minutes at room temperature. Aspirate off the DMEM/F-12 wash from the well and repeat the wash procedure twice more. *Note*: it is very important to perform all the washes to remove any excess substrate prior to visualization.

9) Immediately before the visualization of the fluorescent signal, aspirate off the final wash and add 500 μL of DMEM/F-12 to the well of interest.

10) Visualize the AP activity using a FITC filter under fluorescent microscopy. *Note*: it is very important to perform the visualization step immediately after the final wash, as the fluorescent

(a) (b)

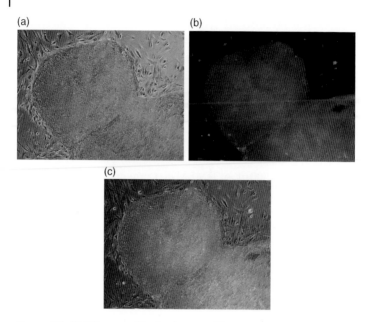

(c)

Figure 4.2 AP 100× magnification images of live staining of established iPSCs grown in feeder-free conditions on Essential 8 medium and Geltrex. PSC colony morphology via phase contrast (a), AP Live Stain (b), and merge of images (a) and (b) (c).

signal is transient; it is cleared from the cells without leaving a footprint within 90 minutes after initial addition.

11) Upon completion of signal visualization and/or manual selection, replace the DMEM/F-12 with normal, prewarmed hPSC culture media and return culture to a humidified incubator, at 37 °C.

12) AP Live stained iPSCs can also be counterstained with the fibroblast marker CD44 signal to distinguish from parental and partially reprogrammed fibroblasts (Figure 4.2).

4.4.2 Immuno-Staining

4.4.2.1 Indirect Immuno-Staining (Two-Step Method) for Self-Renewal Markers

1) Pluripotent self-renewal marker (SSEA-4, TRA-1-60, TRA-1-81) detection can be performed on the original TC dish (12-well plate, scale up or down as needed), as they are surface markers and do not require fixation or permeabilization.

2) Aspirate off the hPSC growth media from the desired sample well of a 12-well plate.

3) Add 1 mL of prewarmed, 37 °C DMEM/F-12 medium and incubate for 2–3 minutes at room temperature.

4) Prepare the appropriate surface marker primary antibody in 0.5 mL of prewarmed 37 °C DMEM/F-12 medium to the recommended dilution.

5) Aspirate off the DMEM/F-12 wash and add the premade primary antibody dilution.

6) Incubate the cultures with the primary antibody at 37 °C for 30–45 minutes in the 37 °C, 5% CO_2, humidified incubator.

7) Following the incubation period, aspirate off the primary antibody solution and add 1 mL of prewarmed DMEM/F-12 medium and incubate at room temperature for 3–5 minutes.

8) Aspirate off the first wash and repeat.

9) Aspirate the second wash and add the AlexaFluor 488 conjugated secondary (or fluorophore of choice) at a 1:500 dilution in DMEM/F-12 and incubate at 37 °C for 30 minutes.

10) Aspirate off the secondary antibody solution and wash twice, as before, with 1 mL prewarmed DMEM/F-12 medium.

11) Right before visualization of the fluorescent signal, aspirate off the second wash and add 1 mL of prewarmed DMEM/F-12 medium. *Note*: prewarmed FluoroBrite DMEM medium may be used to decrease background fluorescence, especially when using fluorophores that excite/emit in the green fluorescence spectrum.

12) Detect the pluripotent marker using a FITC filter (or laser/filter combination optimal for the fluorophore used) on the fluorescent microscope (Figure 4.3).

4.4.2.2 Direct Immuno-Staining (One-Step Method) for Self-Renewal Markers

1) Pluripotent self-renewal marker (SSEA-4, TRA-1-60, TRA-1-81) detection can be performed on the original TC dish (12-well plate, scale up or down as needed), as they are surface markers and do not require fixation or permeabilization. The use of primary antibodies conjugated to fluorophores decreases the time necessary for staining and permits multiplexing, without worrying about cross-reactivity between antibodies.

(a) (b)

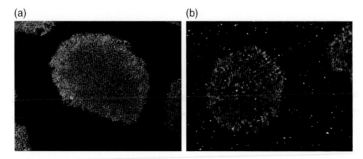

Figure 4.3 Surface antibody staining with antibodies against surface antigens specific to the hPSC self-renewal surface markers SSEA4 (a) and TRA-1-60 (b).

2) Aspirate off the normal PSC growth media from the desired sample well of a 12-well plate.
3) Add 1 mL of prewarmed, 37 °C DMEM/F-12 medium and incubate for 2–3 minutes at room temperature.
4) Prepare the appropriate surface marker antibody in 0.5 mL of prewarmed 37 °C DMEM/F-12 medium to the recommended dilution.
5) Aspirate off the DMEM/F-12 wash and add the prepared dilution of conjugated antibody.
6) Incubate the cultures with the antibody at 37 °C for 30–45 minutes in the 37 °C, 5% CO_2, humidified incubator.
7) Following the incubation period, aspirate off the antibody solution and add 1 mL of prewarmed DMEM/F-12 medium and incubate at room temperature for 3–5 minutes.
8) Aspirate off the first wash and repeat.
9) Right before visualization of the fluorescent signal, aspirate off the second wash and add 1 mL of prewarmed DMEM/F-12 medium. *Note*: prewarmed FluoroBrite DMEM medium may be used to decrease background fluorescence, especially when using fluorophores that excite/emit in the green fluorescence spectrum.
10) Detect the pluripotent marker using a FITC filter (or laser/filter combination optimal for the fluorophore used) on the fluorescent microscope (Figure 4.4).

(a)   (b)

(c)   (d)

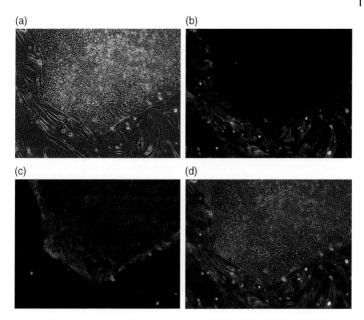

Figure 4.4 iPSC on feeders stained with a positive marker that specifically stains PSCs and not fibroblast feeders (SSEA4) and a negative marker that does not stain PSCs but does stain fibroblast feeders (CD44). Phase contrast (a), CD44-AlexaFluor 488 (b), SSEA4-AlexaFluor647 (c), merge of images (a–c) (d). All images were captured at 40× magnification.

4.4.2.3   Indirect Immunocytochemistry of Intracellular Self-Renewal Marker (Two-Step Method)

1) Pluripotent intracellular self-renewal marker (OCT4, SOX2, NANOG) detection can be performed on the original TC dish (12-well plate, scale up or down as needed). This procedure requires fixation, permeabilization, and blocking prior to proceeding with the immuno-staining (ICC).
2) Aspirate off the normal culture medium from the well to be stained.
3) Add 1 mL of D-PBS and incubate at room temperature for 2–3 minutes.
4) Aspirate off the D-PBS wash.
5) Add 0.5 mL of 4% paraformaldehyde (PFA) to the well and incubate for 15–20 minutes at room temperature.

6) Aspirate off the 4% PFA and add 1 mL of D-PBS and incubate for 5 minutes at room temperature.

7) Aspirate off the D-PBS wash and repeat the wash procedure.

8) Aspirate off the second D-PBS wash and add 1 mL of ICC blocking buffer per well. Incubate for at least 1 hour at room temperature. *Note*: blocking can be performed at 4 °C overnight.

9) Prepare the desired primary antibody dilution in 0.5 mL of blocking buffer; for example, use rabbit anti-OCT4 antibody diluted in blocking buffer 1:500.

10) Aspirate off the blocking solution from the well and add the primary antibody solution and incubate at 4 °C overnight.

11) Following overnight incubation at 4 °C, aspirate off the primary antibody solution.

12) Add 1 mL of D-PBS and incubate at room temperature for 5 minutes.

13) Aspirate off the first D-PBS wash and repeat the wash procedure twice more to remove any unbound primary antibody.

14) Prepare a secondary antibody, conjugated to the fluorophore of your choosing in D-PBS. For example, goat anti-rabbit IgG secondary conjugated to AlexaFluor 594 (1:1000 dilution).

15) Aspirate off the third D-PBS wash from the well and the diluted secondary antibody solution and incubate for 1 hour at room temperature, protected from light.

16) Following the incubation, aspirate off the secondary antibody solution.

17) Add 1 mL of D-PBS and incubate at room temperature for 5 minutes.

18) Aspirate off the first D-PBS wash and repeat the wash procedure once more to remove any unbound secondary antibody.

19) Prepare a nuclear counterstain by adding 2 drops of NucBlue Fixed Cell ReadyProbes Reagent to 1 mL of D-PBS. *Note*: DAPI can also be used to stain the DNA in the nucleus.

20) Aspirate off the second D-PBS wash and add 1 mL of the prepared nuclear counterstain solution. Incubate for 10 minutes at room temperature, protected from light.

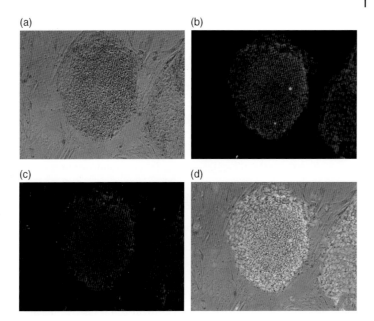

Figure 4.5 Intracellular staining of hPSCs grown on feeders with Oct4, phase contrast (a), OCT4-AlexaFluor 594 (b), counterstained with DAPI (c), and merge of (a–c) (d). All images were captured at 40× magnification.

21) Aspirate off the nuclear counterstain and add 1 mL of D-PBS and incubate at room temperature, protected from light.

22) Aspirate off the D-PBS and add 1 mL of fresh D-PBS prior to visualization. Detect the pluripotent marker using a TRIT-C/Cy3 filter (or laser/filter combination optimal for the fluorophore used) on the fluorescent microscope, and view the counterstain using the DAPI filter (Figure 4.5). *Note*: ProLong gold anti-fade reagent can be used at this time to prevent photobleaching of the fluorescent signal.

4.4.2.4   Surface Antigen Immuno-Staining for Flow Cytometry

4.4.2.4.1   *Harvesting of hPSCs from Feeder-Dependent Cultures*

This protocol describes the harvesting of hPSCs routinely cultured on iMEFs in KSR-based medium. In order to assess the quality of the hPSCs, they must be preferentially harvested away from the iMEFs prior to singularization and subsequent IF staining and flow cytometry analysis. The protocol describes the harvest of a 60 mm dish of PSCs (scale up or down as needed).

1) Aspirate off the normal culture media from the dish of hPSCs to be harvested.
2) Add 5 mL of D-PBS and incubate at room temperature for 2–3 minutes.
3) Aspirate off the D-PBS wash and add 2 mL of prewarmed, 37 °C 1 mg/mL collagenase, type IV (1× solution) and incubate the dish at 37 °C for 30–60 minutes.
4) Stop the incubation when the edges of the colonies start to pull away from the plate.
5) Add 2 mL of prewarmed hPSC medium to the original dish and pipette across the surface of the dish to dislodge any remaining colonies. Transfer this 4 mL suspension to a 15 mL conical tube.
6) Add 2 mL of hPSC medium to collect any remaining cell clusters and add to the 15 mL conical tube. Pipette up and down another 2–3 times to resuspend the colonies.
7) Gravity sediment the colonies to the bottom by allowing the tube to stand at room temperature for 3–5 minutes. *Note*: gravity sedimentation will allow the desired hPSC colonies to settle to the bottom of the tube, while any iMEFs, differentiated cells, dead cells, single cells, and undesirably small colony fragments will not settle and can be aspirated away.
8) Following sedimentation, aspirate off and discard the supernatant, and then gently tap the tube to loosen the cell pellet from the bottom of the tube.
9) Add 1 mL of TrypLE, flick the pellet and place the tube in a water bath for 2–5 minutes.
10) Following incubation, gently pipette up and down to ensure a single cell suspension. Add 10 mL of D-PBS to the cell suspension in the conical tube and pipette up and down several times to dilute the TrypLE.
11) Centrifuge the cell suspension at 200× g for 2 minutes.
12) Aspirate off the supernatant and resuspend the pellet with 2–3 mL of prewarmed medium. Proceed to immuno-staining protocol.

4.4.2.4.2 Harvesting of hPSCs from Feeder-Free Cultures

1) Aspirate off the normal culture media from the plate of hPSCs to be harvested.
2) Add 5 mL of D-PBS and incubate at room temperature for 2–3 minutes.
3) Aspirate off the D-PBS wash and add 2 mL of TrypLE to the hPSCs (1 mL per 35 mm dish, 2 mL per 60 mm dish). Incubate at 37 °C for 3–5 minutes.
4) Using a 5 mL pipette, triturate the cells to achieve a single cell suspension and transfer to a 15 mL conical tube. Wash the remaining cells on the plate with 2 mL of D-PBS and add to the conical tube. Add an additional 10 mL of D-PBS to the cell suspension in the conical tube and pipette up and down several times to dilute the TrypLE.
5) Centrifuge the cells at 200× g for 2 minutes.
6) Aspirate off the supernatant and resuspend the pellet with 2–3 mL of prewarmed medium. Proceed to immuno-staining protocol.

4.4.2.4.3 Direct Immuno-Staining for Flow Cytometry (One-Step Method)

1) Using the resuspended single cell hPSCs, count the cells using the desired method.
2) Using additional DMEM/F-12 medium, dilute out the cells to a concentration of 1×10^6 cells per mL of medium. Aliquot 1 mL of the diluted cells into microcentrifuge tubes for each antibody to be used, making sure to use appropriate unstained controls.
3) For routine hPSC QC per 1×10^6 cells in 1 mL, add 10 μL of SSEA-1 conjugated antibody and 20 μL of TRA-1-60 conjugated antibody (dilution may vary depending upon the antibody and source).
4) Incubate the antibody/cell suspension for 45 minutes at room temperature with constant agitation.
5) Centrifuge the cells at 200× g for 2 minutes.
6) Gently aspirate off the supernatant without disturbing the cell pellet. Add 1 mL of DMEM/F-12 medium to each sample and resuspend the cell pellet by pipetting up and down using a P1000. Incubate for 5 minutes at room temperature.

7) Centrifuge the cells at 200× g for 2 minutes. Aspirate off the wash and repeat the wash another time.

8) Following the second wash step, centrifuge the cells at 200× g for 2 minutes. Aspirate off the wash supernatant and resuspend the cell pellet in 0.5 mL of DPBS (-/-).

9) Filter the single cell suspension through the cell strainer cap of the FACS sample tube.

10) Vortex samples prior to processing the sample through a flow cytometer or the Attune Acoustic Focusing Cytometer.

11) Adjust the settings using the unstained negative control and gate out dead cells, debris, and doublets. Acquire each sample and analyze using FlowJo software.

12) Undifferentiated PSCs are expected to have >85% TRA-1-60 expression and <5% SSEA-1 (differentiation marker) expression (Figure 4.6a) while differentiating EBs show decreased levels of the self-renewal marker TRA-1-60 and increased levels of the differentiation marker SSEA-1 (Figure 4.6b).

4.4.3 Embryoid Body (EB) Formation

This protocol is not for PSCs cultured in flasks. The ability to remove differentiated cells and the availability of cultures to be subjected to this protocol require hPSC cultures to be pristine in normal morphology, with the removal of differentiated cells in order not to skew the trilineage potential. In addition, if using the StemPro EZ Passage disposable stem cell passaging tool and cell scraper, it is best not to grow cells in flasks where they cannot be reached, for selective removal or for overall harvesting.

4.4.3.1 Harvesting Human PSCs Cultured on iMEFs

1) Using a 70% confluent hPSC culture that is in the log phase of replication, aspirate off the spent hPSC medium from the PSC culture dish.

2) Add 2 mL of (1×) collagenase IV solution to a 60 mm dish containing PSCs. Adjust the volume of collagenase IV for various dish sizes, as per Table 4.1.

3) Incubate the PSC cultures for 30–60 minutes in a 37 °C, 5% CO_2 incubator. *Note*: incubation times may vary among different batches of collagenase and different types of PSCs, so you need to determine the appropriate incubation time by

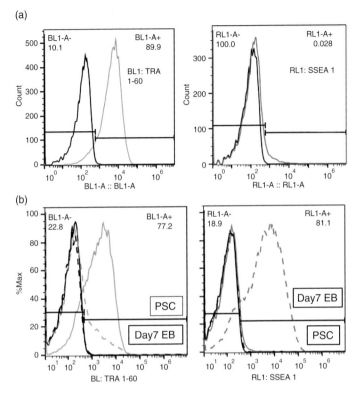

Figure 4.6 Flow cytometry analysis of PSCs. Flow cytometry analysis of undifferentiated PSCs with >85% TRA-1-60 self-renewal marker expression and less than 5% SSEA-1 (differentiation marker) expression (a). Flow cytometry analysis of hPSCs and EBs differentiated from hPSCs demonstrates the loss of self-renewal markers such as TRA-1-60 and the gain of differentiation markers such as SSEA-1 during the differentiation process (b).

Table 4.1 Volume of collagenase IV for PSC enzymatic passaging.

Vessel	6-well plate	35 mm dish	60 mm dish	100 mm dish	T25 flask
Surface area	$10\,cm^2$	$10\,cm^2$	$20\,cm^2$	$60\,cm^2$	$25\,cm^2$
Volume	1 mL	1 mL	2 mL	5 mL	2 mL

examining the colonies. As an alternative, use Dispase at a concentration of 2 mg/mL instead of collagenase IV. Incubate 5–15 minutes in a 37 °C, 5% CO_2 incubator. The EZ Passage tool can also be used in conjunction with a cell scraper as described in Chapter 2.

4) Stop the incubation when the edges of the colonies start to pull away from the plate. Add 2 mL of PSC medium to the culture dish.

5) After incubation, gently dislodge the colonies with a 5 mL pipette by gently pipetting colonies across the surface of the plate 8–12 times. This breaks up the colonies into smaller clumps.

6) Transfer the 4 mL colony suspension to a 15 mL conical tube.

7) Add 2 mL of prewarmed PSC media to the original dish and pipette across the surface of the dish to dislodge any remaining colonies. Transfer this 2 mL suspension to the 15 mL conical tube and pipette up and down 2–3 times to resuspend colonies (be sure not to introduce bubbles or shear colonies too much).

8) Centrifuge the cells at 200× g for 2 minutes at room temperature. Alternatively, you can let the colonies settle to the bottom of the tube via gravity by allowing the tube to stand at room temperature for 5–10 minutes. *Note*: gravity sedimentation will allow desired PSC colonies to settle while any iMEFS, differentiated cells, dead cells, single cells, and undesirably small colony fragments will not settle and can be aspirated off.

9) Aspirate and discard the supernatant, and then gently tap the tube to loosen the cell pellet from the bottom of the tube.

10) Add 5 mL of prewarmed hPSC media and resuspend the PSC fragments by gently pipetting up and down. *Note*: avoid making single cell suspension.

11) Transfer the 5 mL cell cluster suspension into a 60 mm Petri (non TC) dish in a total of 5 mL volume to ensure non-adherent cell culture. *Note*: ultra-low binding plates can be used instead of Petri dishes.

12) Incubate the cell clusters in a 37 °C incubator with 5% CO_2.

13) Following 24 hours of incubation, remove the EBs and media from the Petri dish, transfer the contents to a 15 mL conical

tube and allow the EBs to sediment down via gravity in the cell culture hood at room temperature, for 10–15 minutes.

14) Aspirate off spent medium without disrupting the cell pellet.

15) Replace with 5 mL of fresh EB media and transfer to a fresh 60 mm Petri dish. Place in the incubator and continue to culture.

16) Feed EBs with EB medium every other day for 4 days. Use the same procedure of sedimenting the EBs for media changes. During this time, the cell clusters continue to grow in diameter and take on a spherical shape.

17) For immuno-staining of cells indicative of trilineage potential from random differentiation, proceed to section 4.4.3.1.1. For ScoreCard snalysis, see section 4.4.3.1.2. One plate of EBs can be split and used for both analysis methods (i.e. plate half of the cells onto Geltrex on Day 4 for ICC, and keep the other half in suspension culture until Day 7 for ScoreCard analysis).

4.4.3.1.1 *Option A: Immunocytochemistry (ICC) on Day 21*

1) On Day 4, seed the EBs in EB medium on Geltrex matrix-coated tissue culture dishes and allow the EBs to adhere overnight.

2) Change the spent EB medium every other day for 17 days.

3) On Day 21, perform ICC. Cells can be fixed with 4% paraformaldehyde for further staining (Figure 4.7).

4.4.3.1.2 *Option B: Analysis on Day 7 using the TaqMan hPSC Scorecard Panel*

1) Continue incubation at 37 °C, 5% CO_2 until Day 7, repeating the Day 1 harvest and feed procedure every other day.

2) On Day 7, harvest the EBs for analysis using the TaqMan hPSC Scorecard Panel.

4.4.3.2 Harvesting Human PSCs Cultured Feeder Free in Essential 8 Medium

1) When the cultures are 80–85% confluent, the cells are ready to be harvested for EB formation. *Note*: it is important that the colonies are not small and overcrowded, but rather are allowed to grow robust in individual size for about 4 days (~1200 to 1500 μm in width).

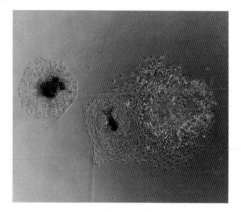

Figure 4.7 Embryoid bodies can be seeded on ECM-coated dishes to facilitate further spontaneous differentiation and migration of cells that can mature into cell phenotypes indicative of the three developmental germ layers.

2) Aspirate the spent medium from the culture vessel and briefly wash once with 5 mL of D-PBS without Ca and Mg (for a 60 mm dish).

3) Aspirate off the D-PBS and add 2 mL of 5× collagenase IV solution (5 mg/mL), prewarmed to 37 °C. Ensure complete coverage of culture surface with the collagenase IV solution.

4) Incubate the cultures grown on vitronectin for 5–15 minutes in a 37 °C, 5% CO_2 incubator until the edges of the colonies begin to curl and detach from the plate. Do not overexpose the cultures to 5× collagenase IV solution. *Note*: cultures grown on Geltrex matrix may take longer to detach. Incubate cultures grown on Geltrex matrix for 15–20 minutes.

5) Aspirate off the 5× collagenase solution and briefly wash the culture with 5 mL of prewarmed DMEM/F-12 medium.

6) Aspirate off the wash solution and add 3 mL of complete PSC medium containing 4 ng/mL of bFGF to dilute any remaining collagenase IV solution.

7) Gently dislodge the colonies from the plate using a cell scraper, and then wash by pipetting them up and down a few times in a 5 mL serological pipette. *Note*: optimal fragment size for the colonies is critical for successful EB formation. Make sure not to triturate the colonies into very small fragments to ensure good fragment size.

8) Transfer the suspended colony clusters into a 15 mL conical tube.

9) Add an additional 2 mL of complete PSC medium to dislodge the remaining colonies and transfer them to the 15 mL tube.

10) Let the colony fragments sediment at the bottom of the 15 mL tube for 5–10 minutes by gravity.

11) Gently aspirate off the supernatant, add 3 mL of complete PSC medium with 4 ng/mL of bFGF, and gently resuspend the sedimented colony fragments by pipetting up and down twice.

12) A total of 5 mL of complete PSC medium is recommended for a 60 mm non-TC treated dish. Transfer the 3 mL cell suspension drop-wise to a non-TC treated culture dish that has been prealiquoted with 2 mL of complete PSC medium with 4 ng/mL of bFGF. This will give a final volume of 5 mL in the 60 mm culture dish.

13) Place the culture dish containing the cell clusters in a 37 °C, 5% CO_2 incubator and incubate overnight.

14) Following overnight incubation, transfer the contents of the non-TC dish to a 15 mL conical tube. Use 2 mL of EB medium to wash the dish to gather any remaining EBs and pool into the conical tube.

15) Gravity sediment the EBs for 5–10 minutes.

16) Aspirate off the supernatant; this step removes bFGF and single cells.

17) Resuspend the sedimented EBs in 5 mL of EB medium.

18) Transfer all 5 mL of EB suspension drop-wise to a new 60 mm non-TC treated dish.

19) Feed EBs with EB medium every other day for 4 days. Use the same procedure of sedimenting the EBs for media changes. During this time, the cell clusters continue to grow in diameter and take on a spherical shape.

20) For immuno-staining of cells indicative of trilineage potential from random differentiation, proceed to section 4.4.3.2.1. For ScoreCard analysis, proceed to section 4.4.3.2.2. One plate of EBs can be split and used for both analysis methods (i.e. plate half of the cells onto Geltrex on Day 4 for ICC, and keep the other half in suspension culture until Day 7 for ScoreCard analysis).

4.4.3.2.1   Option A: Immunocytochemistry (ICC) on Day 21

1) On Day 4, seed the EBs in EB medium on Geltrex matrix-coated tissue culture dishes.
2) Change the spent medium every other day for 17 days.
3) On Day 21, perform ICC. Cells can be fixed with 4% paraformaldehyde for further staining.

4.4.3.2.2   Option B: Analysis on Day 7 Using the TaqMan hPSC Scorecard Panel

1) Continue incubation at 37 °C, 5% CO_2 until Day 7, repeating the Day 1 harvest and feed procedure every other day.
2) On Day 7, harvest the EBs for analysis using the TaqMan hPSC Scorecard Panel.

4.4.3.3   Indirect Immuno-Cytochemistry of Trilineage Differentiation Markers (Two-Step Method)

Embryoid bodies are commonly stained for the three somatic germ lineages: the neuronal marker beta III tubulin for ectoderm, the hepatic marker alpha-fetoprotein (AFP) for endoderm, and the cardiac marker smooth muscle actin (SMA) for mesoderm.

1) Aspirate off the normal culture medium from the well to be stained.
2) Add 1 mL of D-PBS and incubate at room temperature for 2–3 minutes.
3) Aspirate off the D-PBS wash.
4) Add 0.5 mL of 4% paraformaldehyde (PFA) to the well and incubate for 15–20 minutes at room temperature.
5) Aspirate off the 4% PFA and add 1 mL of D-PBS and incubate for 5 minutes at room temperature.
6) Aspirate off the D-PBS wash and repeat the wash procedure.
7) Aspirate of the second D-PBS wash and add 1 mL of ICC blocking buffer per well. Incubate for at least 1 hour at room temperature. *Note*: blocking can be performed at 4 °C overnight.
8) Prepare the desired primary antibody dilution in 0.5 mL of blocking buffer; for example, use mouse anti-AFP antibody diluted in blocking buffer (1:500).

9) Aspirate off the blocking solution from the well and add the primary antibody solution and incubate at 4 °C overnight.

10) Following overnight incubation at 4 °C, aspirate off the primary antibody solution.

11) Add 1 mL of D-PBS and incubate at room temperature for 5 minutes.

12) Aspirate off the first D-PBS wash and repeat the wash procedure twice more to remove any unbound primary antibody.

13) Prepare a secondary antibody, conjugated to the fluorophore of your choosing in D-PBS. For example, goat anti-mouse IgG secondary conjugated to AlexaFluor 594 (1:1000 dilution).

14) Aspirate off the third D-PBS wash from the well and the diluted secondary antibody solution and incubate for 1 hour at room temperature, protected from light.

15) Following the incubation, aspirate off the secondary antibody solution.

16) Add 1 mL of D-PBS and incubate at room temperature for 5 minutes.

17) Aspirate off the first D-PBS wash and repeat the wash procedure once more to remove any unbound secondary antibody.

18) Prepare a nuclear counterstain by adding 2 drops of NucBlue Fixed Cell ReadyProbes Reagent to 1 mL of D-PBS. *Note*: DAPI can also be used to stain the DNA in the nucleus.

19) Aspirate off the second D-PBS wash and add 1 mL of the prepared nuclear counterstain solution. Incubate for 10 minutes at room temperature, protected from light.

20) Aspirate off the nuclear counterstain and add 1 mL of D-PBS and incubate at room temperature, protected from light.

21) Aspirate off the D-PBS and add 1 mL of fresh D-PBS prior to visualization. Detect the pluripotent marker using a TRIT-C/Cy3 filter (or laser/filter combination optimal for the fluorophore used) on the fluorescent microscope, and view the counterstain using the DAPI filter. *Note*: ProLong gold anti-fade reagent can be used at this time to prevent photobleaching of the fluorescent signal (Figure 4.8).

(a)

(b)

(c)

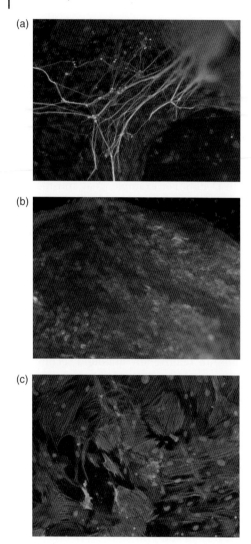

Figure 4.8 hPSCs spontaneously differentiated via EB formation and allowed to mature in culture for 21 days in order to characterize trilineage potential as a measure of pluripotency. Cultures are then probed with antibodies via ICC for cells indicative of the three somatic germ lineages: the neuronal marker beta III tubulin for ectoderm (a), the hepatic marker alpha-fetoprotein (AFP) for endoderm (b), and the cardiac marker smooth muscle actin (SMA) for mesoderm (c). (*See insert for color representation of the figure.*)

4.4.4   TaqMan hPSC Scorecard – From Total RNA

4.4.4.1   Sample Generation
4.4.4.1.1   Generation of Undifferentiated Cells in TRIzol Reagent

1) Cells cultured on iMEF feeders must be cultured under feeder-free conditions on Geltrex or vitronectin matrix-coated culture vessels in iMEF-conditioned medium (CM) or Essential 8 medium for at least one passage. For feeder-free cultures, you may harvest directly from the culture dish.
2) Prior to harvesting for RNA isolation, cells need to be observed under a phase contrast microscope to confirm homogeneous morphology with little or no differentiation.
3) Aspirate off the spent growth medium from the cells and wash once with D-PBS.
4) Add 0.5–1 mL of TRIzol reagent directly on the cells and gently swirl until the liquid is equally distributed across the plate and incubate at room temperature for 3–5 minutes.
5) Use a sterile cell scraper to collect the TRIzol-treated cell slurry.
6) Using a 1 mL pipette, carefully remove all the lysed cells into a sterile DNase/RNase free microcentrifuge tube. Store it at −80 °C until ready for RNA isolation.

4.4.4.1.2   Harvesting Embryoid Bodies in TRIzol Reagent

1) On Day 7 of EB suspension, gently transfer the cells and the medium from the Petri dish into a 15 mL conical tube. Use an additional 5 mL of D-PBS to collect any remaining EBs from the culture dish and add into the conical tube.
2) Allow the EBs to sediment down by gravity for 10–15 minutes, and then aspirate off the supernatant (i.e., spent EB medium).
3) Using a P1000 pipettor, add 1 mL of TRIzol reagent and pipette up and down to assist in properly breaking up the cell clumps.
4) Incubate the EBs for 3–5 minutes. Repeat pipetting and incubation, if the EBs require more time to be lysed. Collect the slurry into a sterile RNase-free microcentrifuge tube. Store at −80 °C until ready for RNA isolation.

4.4.4.2 Sample Preparation
4.4.4.2.1 Total RNA Preparation

1) Incubate the lysate with TRIzol reagent at room temperature for 5 minutes to allow complete dissociation of nucleoprotein complexes.
2) To the TRIzol lysate, add 0.2 mL chloroform per 1 mL of TRIzol reagent and shake the tube vigorously for 15 seconds.
3) Incubate at room temperature for 2–3 minutes and centrifuge at 12 000× g for 15 minutes at 4 °C.
4) Carefully remove the upper aqueous phase and transfer to a new microcentrifuge tube.
5) Add 0.5 mL 100% isopropanol to the aqueous phase per 1 mL of TRIzol reagent; incubate at room temperature for 10 minutes.
6) Centrifuge at 12 000× g for 10 minutes at 4 °C.
7) Carefully remove the supernatant from the RNA pellet and wash with 1 mL of 75% ethanol.
8) Centrifuge the tube at 7500× g for 5 minutes at 4 °C. Discard the supernatant and air dry the RNA pellet for 5–10 minutes.
9) Resuspend the RNA pellet with 20–50 µL RNase-free water.

4.4.4.2.2 DNase Treatment

1) Add 0.1× the isolated RNA volume of 10× DNase I Buffer and 1 µL rDNase I to the RNA in a clean DNase/RNase free microcentrifuge tube, and mix gently.
2) For a 50 µL reaction:

RNA Sample	1–10 µg
10× DNase I Reaction Buffer	5 µL
rDNase I (2 Units)	1 µL
DEPC-treated water to bring reaction to 50 µL	x µL
Total	**50 µL**

3) Incubate the tube at 37 °C for 20–30 minutes.
4) Add the resuspended DNase Inactivation Reagent (typically 0.1 times the RNA volume) and mix well. Incubate 2 minutes at room temperature, mixing occasionally.
5) Centrifuge at 10 000× g for 1.5 minutes and transfer the RNA to a fresh tube.

4.4.4.2.3 *RNA Quantification*

1) Use Nano Drop to quantify the extracted RNA sample. The quality of the RNA is best assessed using absorbance 260/280 ratio, with the recommended value close to 2.0.
2) RNA integrity can be further assessed by running the samples on a 1% agarose gel and assessing the 2:1 ratio of the 28s and 18s RNA bands and the absence of degraded RNA that appears as a small molecular weight smear.
3) If using a Bioanalyzer, a RIN (RNA integrity number) value of higher than 5 may be sufficient, but higher than 8 is ideal for downstream applications.

4.4.4.2.4 *Generate cDNA by Reverse Transcription*

1) Allow the components of the High-capacity cDNA Reverse Transcription Kit with RNase Inhibitor to thaw on ice.
2) Prepare a 2× RT master mix by mixing the components as listed in Table 4.2.
3) Place the 2× RT master mix on ice and mix gently.
4) Prepare the RNA samples by diluting 1 μg total RNA in a total of 225 μL of RNase-free water.
5) Add 225 μL of 2× RT master mix to the diluted RNA and mix well.
6) Aliquot 50 μL of the above RNA plus RT mix in eight vertical wells of a 96-well plate or an eight-strip PCR tube (Figure 4.9).

Table 4.2 Volumes for 2× RT Master Mix.

2× Master Mix	×1	×10 (1 Sample)	×384 (1 × 384w or 4 × 96w plates)
10× RT Buffer	5 μL	50 μL	190 μL
25× dNTP Mix	2 μL	20 μL	76 μL
10× RT Primers	5 μL	50 μL	190 μL
MultiScribe RT	2.5 μL	25 μL	95 μL
RNase Inhibitor	2.5 μL	25 μL	95 μL
Nuclease-free Water	8 μL	80 μL	304 μL
Total	25 μL	250 μL	950 μL

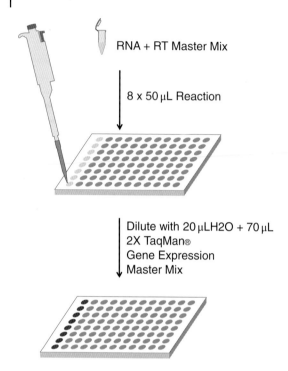

RNA + RT Master Mix

8 x 50 μL Reaction

Dilute with 20 μLH2O + 70 μL
2X TaqMan®
Gene Expression
Master Mix

Figure 4.9 Setting up your cDNA synthesis in eight wells of a 96-well plate or in eight-well PCR strips is highly recommended since it facilitates easy transfer of samples to 96-well and 384-well Scorecard plates. Follow the schematic.

7) Run the RT reaction in a thermal cycler using conditions listed in Table 4.3.
8) Proceed to TaqMan qRT-PCR reaction. If you do not proceed immediately to PCR amplification, store all cDNA samples at −15 °C to −25 °C. To minimize freeze-thaw cycles, store the cDNA in smaller aliquots.

4.4.4.3 Scorecard Assay and Analysis
4.4.4.3.1 *Perform TaqMan qRT-PCR*

1) Dilute each well containing 50 μL cDNA with 20 μL PCR water for a final volume of 70 μL.
2) Add 70 μL 2× TaqMan Gene Expression Master Mix (if using the TaqMan hPSC Scorecard Panel 384w) or 70 μL 2×

Table 4.3 Thermal Cycler condition for RT reaction.

Step	Temperature	Time
1	25 °C	10 minutes
2	37 °C	120 minutes
3	85 °C	5 minutes
4	4 °C	Hold

TaqMan Fast Advanced Master Mix (if using the TaqMan hPSC Scorecard Panel 96w FAST).

3) Spin down the plates at 600× g for 2 minutes.

4) Load 10 μL per well using a multichannel pipette onto the 384-well or 96-well plate using fresh tips each time. For 96-well plates, one well is sufficient to load one row of the plate (Figure 4.10).

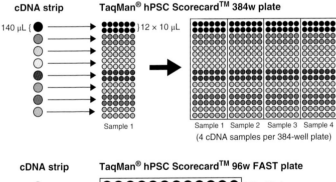

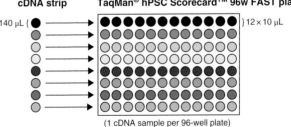

Figure 4.10 Sample set up for cDNA synthesis. Top figure is for the 384-well Scorecard; bottom figure is for the 96-well FAST Scorecard.

Table 4.4 TaqMan ScoreCard cycling parameters.

TaqMan hPSC Scorecard 384w Run mode (Ramp rate): Standard			
Step	Temperature	Time	Cycles
Hold	50 °C	2 minutes	-
Hold	95 °C	10 minutes	-
Melt	95 °C	15 seconds	40
Anneal/Extend	60 °C	1 minute	

5) Seal the plate with MicroAmp Optical Adhesive Film, and centrifuge it at 600× g for 2 minutes.
6) Place the plate in a compatible RT-PCR instrument equipped with the appropriate thermal block.
7) Open the experiment template file and save a separate copy with your experimental details. Run the experiment using Standard method for 384-well plates with the TaqMan Gene Expression Master Mix and Fast mode for 96-well plates with the TaqMan Fast Advanced Master Mix, using the cycling parameters listed in Table 4.4.

4.4.4.3.2 *Analyze Data Using the hPSC Scorecard Analysis Software*

Analyze the gene expression data from the TaqMan hPSC Scorecard Panels using the web-based hPSC Scorecard Analysis Software, available at https://apps.thermofisher.com/hPSCscorecard/home.htm. The hPSC Scorecard Analysis Software summarizes all key experimental results, including pluripotency and differentiation potential, on a single dashboard. It also allows you to tag and filter experiments, view expression, correlation, box plots, heat maps, scores, and export experimental results and data as a pdf or a spreadsheet.

References

1 C. Bock *et al.* Reference maps of human ES and iPS cell variation enable high-throughput characterization of pluripotent cell lines. *Cell* **144**, 439–452 (2011).

2 L. Cheng *et al.* Low incidence of DNA sequence variation in human induced pluripotent stem cells generated by nonintegrating plasmid expression. *Cell Stem Cell* **10**, 337–344 (2012).

3 J.S. Draper *et al.* Recurrent gain of chromosomes 17q and 12 in cultured human embryonic stem cells. *Nat Biotechnol* **22**, 53–54 (2004).

4 T. Enver *et al.* Cellular differentiation hierarchies in normal and culture-adapted human embryonic stem cells. *Hum Mol Genet* **14**, 3129–3140 (2005).

5 A.M. Vitale *et al.* Variability in the generation of induced pluripotent stem cells: importance for disease modeling. *Stem Cells Transl Med* **1**, 641–650 (2012).

6 M. Marti *et al.* Characterization of pluripotent stem cells. *Nat Protoc* **8**, 223–253 (2013).

7 International Stem Cell Initiative. Characterization of human embryonic stem cell lines by the International Stem Cell Initiative. *Nat Biotechnol* **25**, 803–816 (2007).

8 U. Singh *et al.* Novel live alkaline phosphatase substrate for identification of pluripotent stem cells. *Stem Cell Rev* **8**, 1021–1029 (2012).

9 F.J. Muller, B. Brandl, J.F. Loring. Assessment of human pluripotent stem cells with Pluritest. *StemBook*. Cambridge, MA: Harvard Stem Cell Institute, 2008.

10 F.J. Muller *et al.* A bioinformatic assay for pluripotency in human cells. *Nat Methods* **8**, 315–317 (2011).

11 R.H. Quintanilla Jr, J.S. Asprer, C. Vaz, V. Tanavde, U. Lakshmipathy. CD44 is a negative cell surface marker for pluripotent stem cell identification during human fibroblast reprogramming. *PLoS One* **9**, e85419 (2014).

12 R.H. Quintanilla Jr, J. Asprer, K. Sylakowski, U. Lakshmipathy. Kinetic measurement and real time visualization of somatic reprogramming. *J Vis Exp* **113**, doi: 10.3791/54190 (2016).

13 J. Fergus, R. Quintanilla, U. Lakshmipathy. Characterizing pluripotent stem cells Using the TaqMan(R) hPSC Scorecard(TM) Panel. *Methods Mol Biol* **1307**, 25–37 (2016).

5

Differentiation

5.1 Introduction

The ability to generate different types of functional cells from pluripotent stem cells (PSCs) is well established [1]. Most of the successful approaches apply the principles of developmental biology to stem cell differentiation [2]. This comprises stage-specific addition of growth factors, cytokines or small molecules that activate specific signaling pathways in a precise manner. Here, three protocols are provided for the induction of pluripotent stem cells into representative cell stypes of the three germ layers – endoderm, ectoderm, and mesoderm. The first one is a definitive endoderm differentiation that serves as an intermediate for further differentiation to terminal endodermal cells such as hepatic or pancreatic lineages. The second protocol is specific for neural induction for the generation of neural stem cells that can be further expanded and differentiated into neural subcell types [3]. The third protocol is for generation of beating cardiomyocytes from PSCs with methods for further enriching of cardiomyocytes using metabolic enrichment [4].

5.2 Materials

All materials are from Thermo Fisher Scientific unless specified otherwise.

Human Pluripotent Stem Cells: A Practical Guide, First Edition. Uma Lakshmipathy, Chad C. MacArthur, Mahalakshmi Sridharan and Rene H. Quintanilla.
© 2018 John Wiley & Sons, Inc. Published 2018 by John Wiley & Sons, Inc.

5.2.1 Definitive Endoderm Differentiation

1) PSC Definitive Endoderm Induction Kit *Cat# A3062601*
2) StemPro Accutase Cell Dissociation Reagent
3) RevitaCell Supplement or ROCK Inhibitor Y-27632
4) Dulbecco's Phosphate Buffered Saline (D-PBS) without calcium chloride or magnesium chloride

5.2.2 Neural Differentiation

1) PSC Neural Induction Medium *Cat# A1647801*
2) Essential 8™ Medium *Cat# A1517001*
3) Geltrex™ LDEV-Free hESC-qualified Reduced Growth Factor Basement Membrane Matrix *Cat# A1413302*
4) Vitronectin (VTN-N) Recombinant Human Protein, Truncated *Cat# A14700*
5) RevitaCell™ Supplement (100×), Life Technologies *Cat# A26445-01*
6) UltraPure™ 0.5M EDTA, pH 8.0 *Cat# 15575020*
7) Versene Solution *Cat# 15040066*
8) TrypLE™ Select Cell Dissociation Reagent *Cat# 12563*
9) ROCK Inhibitor Y27632, Sigma-Aldrich *Cat# Y0503*
10) Dulbecco's Phosphate Buffered Saline (DPBS) without calcium and magnesium *Cat# 14190-144*
11) StemPro™ Accutase™ Cell Dissociation Reagent *Cat# A1110*
12) Advanced DMEM/F-12 Medium *Cat# 12634010*
13) Neurobasal™ Medium *Cat# 21103049*

5.2.3 NSC Staining

1) Human Neural Stem Cell Immunocytochemistry Kit *Cat# A24354*
   (Nestin, PAX6, SOX1, and SOX2)
2) (*Optional*) ProLong Gold Anti-fade Reagent *Cat# P36930*

5.2.4 Cardiomyocyte Differentiation

1) Dulbecco's Phosphate Buffered Saline (DPBS) without calcium and magnesium *Cat# 14190-144*

2) Essential 8™ Medium *Cat# A1517001*
3) Geltrex™ LDEV-Free hESC-qualified Reduced Growth Factor Basement Membrane Matrix *Cat# A1413302*
4) Vitronectin (VTN-N) Recombinant Human Protein, Truncated *Cat# A14700*
5) RevitaCell™ Supplement (100×), Life Technologies *Cat# A26445-01*
6) TrypLE™ Select Cell Dissociation Reagent *Cat# 12563*
7) PSC Cardiomyocyte Differentiation Kit *Cat# A2921201*
8) RPMI 1640 without glucose *Cat# 11879-020*
9) 25% Human Serum Albumin, Lee Biosolutions *Cat# 101-15-50*
10) L-Ascorbic Acid 2-Phosphate Trisalt, Wako Chemicals *Cat# 323-44822*
11) Sodium DL-Lactate, Sigma *Cat# L4263-100ML*
12) 1M sodium HEPES, Sigma *Cat# H3662-500ML*
13) UltraPure™ DNase/RNase-Free Distilled Water *Cat# 10977-023*
14) Human Cardiomyocyte Immunocytochemistry Kit *Cat# A25973*

5.2.5   TNNT2 Flow Cytometry of Cardiomyocytes

1) TNNT2 antibody, Human Ms IgG1 MAb *Cat# MS295-P1*
2) Goat anti-Mouse IgG1 Secondary Antibody, AlexaFluor™ 488 conjugate *Cat# A21121*
3) Dulbecco's Phosphate Buffered Saline (DPBS) without calcium and magnesium *Cat# 14190-144*
4) Fetal Bovine Serum (FBS) ES Cell-qualified *Cat# 16141-061*
5) RPMI 1640 without glucose *Cat# 11879-020*
6) Blocker™ BSA (10×) in PBS *Cat# 37525*
7) Paraformaldehyde, 4% Affymetrix *Cat# 19943*
8) 90% methanol: 10 mL distilled water in 90 mL methanol, Fisher Scientific *Cat# A412-1*

5.3   Solutions

5.3.1   PSC Neural Induction Medium (for 500 mL)

Neural Induction Supplement	10 mL
Neurobasal Medium	400 mL

Complete PSC Neural Induction Medium can be stored at 2–8 °C in the dark for up to 2 weeks. Warm the Neural Induction Medium in a 37 °C water bath for 5–10 minutes before using.

5.3.2 Neural Expansion Medium (for 500 mL)

Neural Induction Supplement	10 mL
Advanced DMEM/F-12 Medium	245 mL
Neurobasal Medium	245 mL

Complete PSC Neural Induction Medium can be stored at 2–8 °C in the dark for up to 2 weeks. Warm the Neural Induction Medium in a 37 °C water bath for 5–10 minutes before using.

5.3.3 Cardiac Enrichment Medium: Glucose-Free Medium with Lactate (100 mL)

25% Human Serum Albumin	1 mL
*250× Ascorbic Acid Stock	400 µL
**1 M Lactate Stock	400 µL
RPMI 1640 (glucose-free)	Add up to 100 mL

Sterilize through a 0.22 µm filter unit and store at 4 °C for up to 2 weeks.

5.3.4 *160 mg/mL Ascorbic Acid Stock (250×)

1) Dissolve 1600 mg of ascorbic acid 2-phosphate powder in distilled water to a 10 mL volume. Aliquot and store at −20 °C in the dark for up to 6 months. Once the aliquot is thawed, used immediately and discard unused materials; do not store at 4 °C. *Note*: the powder is difficult to dissolve. Put ~6 mL of the distilled water out of the 10 mL final volume in a 50 mL conical tube and slowly add the powder with periodic mixing. Then bring the volume up to 10 mL. This might still require a lot of vortexing afterwards.
2) When fresh out of the 4 °C fridge, the stock solution tends to be viscous and the ascorbic acid seems incompletely

dissolved. Vortexing and warming to room temperature help dissolve the ascorbic acid again. When adding this solution to the basal medium, make sure to add it very slowly. Move the pipette around to help dissolve the solution as it is added to prevent a glob of undissolved ascorbic acid stuck at the bottom of the filter.

5.3.5   **1 M Lactate Stock

Mix 1.426 mL of commercially available 60% sodium DL-lactate solution with 8.574 mL of 1 M Na-HEPES for a 10 mL solution. Store at 4 °C in the dark, for up to 6 months.

5.3.6   RP20 Buffer (20% FBS) (for 500 mL)

ES-Qualified FBS	100 mL
RPMI 1640 (glucose-free)	400 mL

Sterilize through 0.22 μm filter. Medium lasts for up to 1 month at 4 °C.

5.3.7   Flow Buffer (0.5% BSA) (for 500 mL)

10% BSA Solution	25 mL
D-PBS (-/-)	475 mL

Sterilize through 0.22 μm filter. Medium lasts for up to 1 month at 4 °C.

5.4   Methods

5.4.1   Directed Differentiation of Definitive Endoderm from Human PSCs

This protocol is intended for the harvesting of a six-well dish of feeder-free hPSC grown on Geltrex or vitronectin with Essential 8 medium and then replating those cells into a six-well plate for differentiation (scale up or down as necessary). All media in the

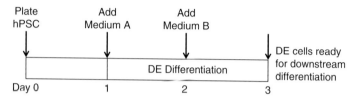

Figure 5.1 Workflow schematic for induction of definitive endoderm differentiation from monolayer of pluripotent stem cells.

PSC Definitive Endoderm Induction Kit should be aliquoted and allowed to come up to room temperature before use; never prewarm the media using a water bath or incubator. The entire workflow is summarized in Figure 5.1.

5.4.1.1   Day 0: hPSC Seeding

1) Coat dishes with matrix (vitronectin or Geltrex).
2) Prewarm required amount of complete Essential 8 medium and Accutase to room temperature until no longer cool to the touch (do not use 37 °C water bath).
3) Wash the cells with 2 mL of room temperature D-PBS (-/-) per well to be harvested.
4) Aspirate off the D-PBS wash.
5) Add 1 mL of prewarmed Accutase per well.
6) Incubate the cells at room temperature for no more than 5 minutes. Stop the incubation as soon as clumps detach with a simple finger tap to the well.
7) Remove the cells by gently squirting the colonies from the well using a 5 mL pipette. Collect cells in a 15 mL conical tube.
8) Wash the wells with 1 mL per well using prewarmed Essential 8 medium and pool with initial cells that were collected.
9) Dilute the cells with (10 times by volume of Accutase used) 10 mL prewarmed Essential 8 medium.
10) Centrifuge the cells at 200× g for 5 minutes at room temperature.
11) Aspirate off the supernatant and gently flick the bottom of tube to loosen the cell pellet.
12) Resuspend the cell pellet in 1 mL prewarmed complete Essential 8 medium and triturate the cells 2–3 times with a 10 mL pipette to create small cell clumps.

13) Dilute the cell clumps to a final 1:10 dilution using 9 mL of prewarmed complete Essential 8 medium containing 5 μM ROCK inhibitor or 1× RevitaCell supplement.
14) Add 2 mL of the cell clump suspension per well of a six-well plate precoated with vitronectin.
15) Shake the plate in a cross pattern to ensure even distribution of the small cell clumps.
16) Place the plate in a 37 °C, 5% CO_2 humidified incubator overnight.

5.4.1.2 Day 1: DE Induction with Medium A

17) Prewarm DE Induction Medium A at room temperature until no longer cool to the touch (do not use 37 °C water bath).
18) Aspirate off spent complete Essential 8 medium.
19) Add 2 mL per well prewarmed DE Induction Medium A.
20) Incubate the plate in a 37 °C, 5% CO_2 incubator for 24 hours.

5.4.1.3 Day 2: DE Induction with Medium B

21) Prewarm DE Induction Medium B at room temperature until no longer cool to the touch (do not use 37 °C water bath).
22) Aspirate off the spent DE Induction Medium A.
23) Add 2 mL per well prewarmed DE Induction Medium B.
24) Incubate the plate in a 37 °C, 5% CO_2 incubator for 24 hours.

5.4.1.4 Day 3: PSCs are Induced

At this point, the PSCs have been induced into definitive endoderm and can be further differentiated into downstream lineages such as midgut/hindgut (expressing CDX2), pancreatic endoderm (expressing PDX2), and liver bud progenitors (expressing AFP). Cells at the definitive endoderm stage can be analyzed via ICC for FoxA2 (Figure 5.2a) or flow cytometry for high expressing of SOX17 (Figure 5.2b).

5.4.2 Directed Differentiation to Neural from Human PSCs

This protocol is monolayer neural induction of feeder-free hPSCs grown on Geltrex or vitronectin with Essential 8 medium into neural stem cells using the PSC Neural induction Medium over a 7-day period, without the need to create embryoid bodies

(a)

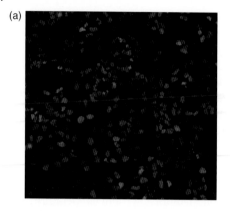

(b)

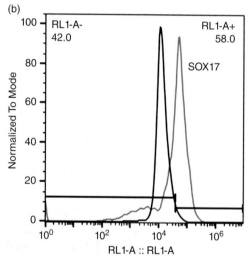

Figure 5.2 Definitive endoderm cells derived from iPSCs hPSCs induced to definitive endoderm, at Day 3. ICC performed using FOXA2 (*red*), counterstained for nuclei (*blue*) with DAPI (a). Flow analysis for Sox17 staining of PSC (*black*) and PSC-derived definitive endoderm (*red*) (b). (*See insert for color representation of the figure.*)

or neurospheres (Figure 5.3). The protocol starts from a 60 mm dish of feeder-free hPSCs and then replating those cells into a six-well plate for differentiation (scale up or down as necessary). NSC induction is complete by Day 7 and the resulting cells can be further expanded using the Neural Expansion medium for

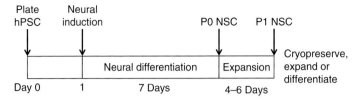

Figure 5.3 Workflow schematic for induction of neural stem cells from monolayer of pluripotent stem cells.

further downstream application. The Human Neural Stem Cell Immunocytochemistry kit can be used to confirm expression of NSC markers post 7 days of induction.

5.4.2.1 Day 0: hPSC Seeding

1) Start with high-quality human PSCs (with minimal or no differentiated colonies) cultured in feeder-free conditions with Essential 8 medium on vitronectin or Geltrex-coated dishes.
2) Coat six-well plates with the same coating matrix on which your PSCs are cultured.
3) When the PSCs reach ~70–80% confluency, dislodge PSCs to generate cell clumps for passaging following the normal protocol.
4) Generate a PSC cell suspension, then transfer a portion of the cell suspension to a 15 mL conical tube (for example, transfer 1 mL of a 6 mL PSC suspension prepared from one well of a six-well plate) to estimate the total cell number of the PSC cell suspension.
5) Centrifuge the 15 mL conical tube with the cells at 200× g for 2 minutes and aspirate off the supernatant.
6) Add 1 mL of prewarmed StemPro Accutase Cell Dissociation Reagent to the 15 mL conical tube containing the cells, and incubate for 5 minutes at 37 °C.
7) Vigorously pipette the cells up and down with a 1 mL pipette five times to dissociate the cells into a single cell suspension and determine the cell density using a hemacytometer or the Countess II automated cell counter.
8) Add 2.5 mL Essential 8 medium into each well of the freshly coated six-well plates.
9) Gently shake the conical tube containing the remaining PSC clusters to get a good cell suspension (the remaining 5 mL).

10) Using the cell concentration determined from the single cell suspension, determine the right amount of cell cluster suspension to achieve a relative seeding density of 2×10^5 to 3×10^5 PSCs per well. For example, add 0.25–0.3 mL of PSC suspension to each well if the concentration of PSC suspension is 1×10^6 cells/ mL.

11) Move the plates in several quick back-and-forth and side-to-side motions to disperse the cells across the surface, then gently place the plates in a 37 °C, 5% CO_2 humidified incubator overnight. *Note*: The split ratio varies depending on the confluence of PSCs before splitting and the variability between PSC lines. Neural induction starts on Day 1 of PSC splitting (about 24 hours after passaging). The starting density of PSCs should be about 15–25% confluency.

12) When passaging PSCs, cells should be plated as small clumps and not as a single cell suspension. Avoid plating PSC as single cells as that can lead to increased cell death. *Note*: to prevent cell death, you may treat the cells overnight with $10 \, \mu M$ of ROCK inhibitor Y-27632 or RevitaCell Supplement by adding it to the Essential 8 medium at the time of splitting.

5.4.2.2 Day 1: Start of Neural Induction

13) PSCs must be at 15–25% confluency prior to starting neural induction. Aspirate off the spent Essential 8 medium to remove non-attached cells.

14) Add 2.5 mL prewarmed complete PSC Neural Induction Medium to each well of the six-well plates. Return the plates to the 37 °C, 5% CO_2 humidified incubator for overnight incubation.

5.4.2.3 Day 2: Neural Induction Observation

15) Confirm that the morphology of cell colonies is uniform; if there are areas in the dish exhibiting bad morphologies, they must be mechanically removed prior to continuing. *Note*: if there are a large number of such non-neural colonies, it is recommended to discard the cultures and start again with high-quality PSCs.

16) About 48 hours after switching to the PSC Neural Induction Medium, aspirate off the spent medium.

17) Add 2.5 mL of prewarmed complete Neural Induction Medium per well.

5.4.2.4 Day 4: Neural Induction Observation

18) On day 4 of neural induction, cells will be reaching confluency. Mechanically remove any unwanted colonies with a Pasteur glass pipette.
19) Aspirate off the spent medium from each well and add 5 mL of prewarmed complete PSC Neural Induction Medium per well.

5.4.2.5 Day 6: Neural Induction Observation

20) On day 6 of neural induction, cells should be near maximal confluence. Remove any non-neural differentiated cells.
21) Aspirate off the spent medium and add 5 mL of complete PSC Neural Induction Medium into each well. *Note*: due to high cell density in the culture from Day 4 onwards, doubling the volume of PSC Neural Induction Medium is very critical for cell nutrition. Also, minimal cell death should be observed from Days 4–7 after neural induction. If the color of cells turns brownish with many floating cells during Days 4–7 of neural induction, this indicates that the starting density of PSCs was too high. In this case, change the Neural Induction Medium every day using 5 mL per well.

5.4.2.6 Day 7: Characterization and Expansion of Neural Stem Cells

22) On Day 7 of neural induction, the resulting cells can be characterized via immunocytochemistry using the Human Neural Stem Cell Immunocytochemistry kit or your antibodies of choice for visualization (Figure 5.4a) or flow analysis (Figure 5.4b). Cells can also be expanded, cryopreserved or further differentiated to the cell type of interest.

5.4.2.7 Harvesting and Expansion of P0 NSCs

1) On Day 7 of neural induction, the resulting NSCs (P0) are ready to be harvested and expanded.
2) Prepare fresh Geltrex-coated vessels prior to usage.
3) Aspirate off the spent Neural Induction Medium and gently add 2 mL of room-temperature D-PBS (-/-) into each well of a six-well plate. *Note*: add the D-PBS towards the wall of the well to avoid cell detachment.

(a)

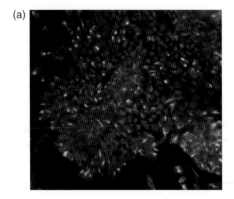

(b)

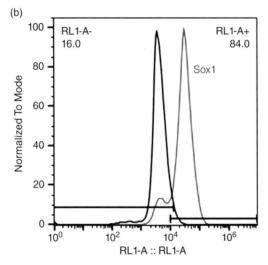

Figure 5.4 Neural stem cells derived from PSCs. hPSCs induced to NSCs. ICC performed using Nestin (*green*), Sox2 (*red*), counterstained for nuclei (*blue*) with DAPI (a). Flow analysis of Sox1 staining of PSCs (*black*) and PSC-derived NSCs (*red*) (b). (*See insert for color representation of the figure.*)

4) Aspirate off the D-PBS wash and gently add 1 mL of pre-warmed StemPro Accutase Cell Dissociation Reagent to each well of the six-well plate. Incubate for 5–8 minutes at 37 °C until most of the cells detach from the surface of the culture vessels.

5) Use a cell scraper to detach the cells off the surface of the plates.

6) Transfer the cell clumps using a 5 mL pipette into a 15 mL conical tube.

7) Add 1 mL of D-PBS to each well of the six-well plate to collect residual cells and transfer the cell suspension to the conical tube.

8) Gently pipette the cell suspension up and down three times with a 5 mL or 10 mL pipette to break the cell clumps.

9) Pass cell suspension through a 100 μm strainer and centrifuge the cells at 300× g for 4 minutes at room temperature.

10) Aspirate off the supernatant, resuspend the cells with D-PBS (3–5 mL of D-PBS for all cells from one well of a six-well plate) and centrifuge the cells at 300× g for 4 minutes at room temperature.

11) Aspirate off the supernatant and resuspend the cells with prewarmed complete Neural Expansion Medium (1 mL for all cells from one well of a six-well plate).

12) Determine the cell concentration using your preferred method.

13) Dilute the cell suspension with prewarmed complete Neural Expansion Medium to 2×10^5 to 4×10^5 cells/ mL.

14) Add ROCK inhibitor Y27632 to a final concentration of 5 μM, or RevitaCell supplement (1×) into the medium used to suspend and plate the cells.

15) Add the diluted cell suspension into each freshly coated Geltrex culture plate/dish to plate the cells at a density of 0.5×10^5 to 1×10^5 cells/cm^2.

16) Move the culture vessels in several quick back-and-forth and side-to-side motions to disperse the cells across the surface and place them gently in the incubator. *Note*: avoid splashing the medium to the outsides of the well to avoid contamination.

17) After overnight incubation, aspirate off the spent complete Neural Expansion Medium to eliminate the ROCK inhibitor. Add the appropriate amount of Neural Expansion Medium to each vessel as appropriate to the size. Continue to exchange the Neural Expansion Medium without Y27632 every other day thereafter.

18) Usually, NSCs reach confluency on Day 4–6 after plating. When NSCs reach confluency, they can be further expanded in complete Neural Expansion Medium. Expanded NSCs

can be cryopreserved or differentiated into specific neural cell types following the protocol of your choice. *Note*: after dissociation of P0–P4 NSCs, the overnight treatment with the ROCK inhibitor Y27632 at a final concentration of 5 µM or RevitaCell supplement is required at the time of plating to prevent cell death for both expansion and differentiation into glial and neuronal cells.

5.4.3 Directed Differentiation of Cardiomyocyte from Human PSCs

This protocol describes generation of cardiomyocytes from human ESCs/iPSCs cultured in feeder-free conditions on Geltrex and Essential 8 medium, which are then differentiated utilizing a commercially available PSC Cardiomyocyte Differentiation Kit. The overall protocol is highlighted in Figure 5.5. Cardiomyocyte differentiation from pluripotent stem cells is carried out using the PSC Cardiomyocyte Differentiation Kit, which consists of sequential media transitions to induce differentiation. For most PSC lines, the TNNT2+ cardiomyocyte-positive population exceeds 60–70% of the total population. In such instances, an enrichment step is optional. Enrichment of PSC-derived cardiomyocytes using glucose-free lactate medium permits the enrichmentofcardiomyocytesbyleveragingtheswitchfromglucose-based metabolism to lactate metabolism, specific for cardiomyocytes.

5.4.3.1 Cardiomyocyte Differentiation from Pluripotent Stem Cells

This protocol is intended for the harvesting of a 60 mm dish of feeder-free PSCs grown on Geltrex with Essential 8 medium and then replating those cells into a six-well plate for

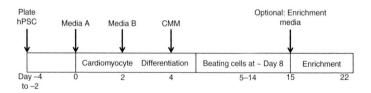

Figure 5.5 Workflow schematic for induction of cardiomyocytes from monolayer of pluripotent stem cells.

differentiation (scale up or down as necessary). All media in the PSC Cardiomyocyte Differentiation Kit should be aliquoted and allowed to come up to room temperature before use; never prewarm the media by using a water bath or incubator.

5.4.3.1.1 Day –4: Harvesting and Seeding of PSCs

1) Aspirate off the Essential 8 growth medium from the PSC culture and wash dish once with 5 mL D-PBS (-/-) at room temperature for 2 3 minutes.
2) Aspirate off the D-PBS wash and add 2 mL of TrypLE Express reagent and ensure even distribution to the whole plate surface. Incubate the cells at 37 °C for 3–5 minutes or until the cells begin to round up and separate from colony formation and can be readily detached.
3) Triturate the cells 3–5 times by rinsing the cell suspension across the surface of the dish.
4) Transfer the cell suspension to a sterile 15 mL conical tube. Add 2 mL of Essential 8 medium to the dish and rinse off the remaining cells and pool the cells into the conical tube. Add an additional 6 mL of Essential 8 medium to the conical tube and mix the cell suspension to dilute out the TrypLE.
5) Centrifuge the cell suspension at 200× g for 2 minutes at room temperature. Carefully aspirate off the supernatant without disturbing the cell pellet.
6) Gently flick the tube 3–5 times to loosen the cell pellet, and resuspend pellet in 2–3 mL of Essential 8 medium containing 1× RevitaCell supplement.
7) Determine the cell concentration and percent viability using a hemacytometer or the Countess II automated cell counter or similar device/method. *Note*: cell viability is typically >90%.
8) For plating PSCs for differentiation onto Geltrex-coated six-well TC dishes, a range of seeding densities are recommended per well of the six-well plate (10 cm^2 surface area per well) in 2 mL of Essential 8 medium per recommendation. *Note*: different seeding densities must be used for each cell line to determine the optimal concentration necessary to achieve optimal cardiomyocyte differentiation. It is recommended to start with the following seeding densities to determine the optimal density: 5000 cells/cm^2, 10 000 cells/cm^2 and 20 000 cells/cm^2, for initial testing of new lines.

5.4.3.1.2 Day −3 to 0: Expand PSCs Culture

9) Aspirate off the spent medium and add 2 mL of fresh pre-warmed (room temperature) Essential 8 medium into each well of a six-well plate (RevitaCell is not required 24 hours post seeding).

10) Incubate the cultures overnight in a 37 °C, 5% CO_2 incubator. *Note*: continue to culture the PSCs in Essential 8 medium until the cultures have reached 65–80% confluency. (Automated imaging systems such as the IncuCyte Zoom can be used to measure the confluence starting 24 hours post seeding to measure growth rates and confluence.)

5.4.3.1.3 Day 0: Change Medium to Cardiomyocyte Differentiation Medium A

Note: the PSC culture should exhibit 65-80% confluence. Shedding of dead cells is normal.

11) Aspirate off the spent Essential 8 medium and replace with 2 mL/well Cardiomyocyte Differentiation Medium A, prewarmed to room temperature.

12) Return the cultures to the 37 °C incubator with a humidified atmosphere of 5% CO_2.

5.4.3.1.4 Day 2: Change Medium to Cardiomyocyte Differentiation Medium B

Note: the cells will start to become more opaque. Shedding of dead cells is normal.

13) Aspirate off the spent Medium A and replace with 2 mL/well Cardiomyocyte Differentiation Medium B, prewarmed to room temperature.

14) Return the cultures to the 37 °C incubator with a humidified atmosphere of 5% CO_2.

5.4.3.1.5 Day 4: Change Medium to Cardiomyocyte Maintenance Medium (CMM)

15) Aspirate the spent Medium B, and replace with 2 mL/well of Cardiomyocyte Maintenance Medium, prewarmed to room temperature.

16) Return the cultures to the 37 °C incubator with a humidified atmosphere of 5% CO_2.

5.4.3.1.6 Day 6–10: Refeed Every Other Day with Cardiomyocyte Maintenance Medium

Note: on days 6 and 8, refeed differentiating cells with CMM. Contracting cardiomyocytes can be observed as early as Day 7 post differentiation induction.

17) Aspirate the spent CMM and replace with 2 mL/well of CMM, prewarmed to room temperature.
18) Return the cultures to the 37 °C incubator with a humidified atmosphere of 5% CO_2.

5.4.3.1.7 Day 10: Characterize Cardiomyocyte

19) On Day 10, spontaneously contracting syncytium of troponin T cardiac type 2 (TNNT2/cTnT) positive cardiomyocytes will be present for analysis or further applications.
20) Analyze via immunocytochemistry (ICC). The Human Cardiomyocyte Immunocytochemistry Kit can be used to stain for two common human cardiac lineage markers: NKX2-5 for early cardiac mesoderm and TNNT2/cTNT for cardiomyocytes (Figure 5.6a). For analysis via flow cytometry, the TNNT2/cTNT antibody from the kit or another TNNT2 primary antibody can be used in conjunction with the appropriate secondary conjugated to a fluorophore (Figure 5.6b). For enrichment of cardiomyocyte cultures, users can follow the enrichment protocol for cardiomyocytes as early as 7 days after the first contracting cardiomyocytes start to appear.

5.4.3.2 Enrichment of hPSC-Derived Cardiomyocytes

This protocol is meant to be used in conjunction with the cardiomyocytes that have been created using the PSC Cardiomyocyte Differentiation Kit. This is a 7-day protocol, which begins about 1 week after contracting cardiomyocytes start to appear with the PSC Cardiomyocyte Differentiation Kit. Cardiomyocyte differentiation is recommended in six-well format if intending to enrich or analyze via flow cytometry, or for use in further downstream applications.

1) Starting with a healthy culture of contracting PSC-derived cardiomyocytes, aspirate off the spent Cardiomyocyte Maintenance Medium from the wells of the six-well plate to be enriched.
2) Add 2 mL/well prewarmed enrichment medium.

(a)

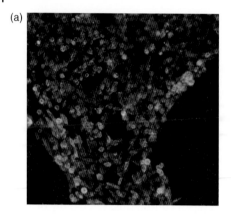

(b)

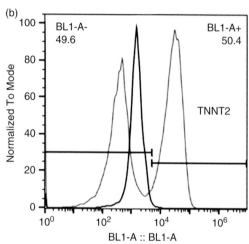

Figure 5.6 Cardiomyocytes derived from PSCs. hPSCs induced to cardiomycotyes. ICC performed using TNNT2 (*green*) and NKX2.5 (*red*) counterstained for nuclei (*blue*) with DAPI (a). Flow analysis of TNNT2 staining of PSCs (*black*) and PSC-derived cardiomyocytes (*green*) (b). (*See insert for color representation of the figure.*)

3) Return the cultures to the 37 °C incubator with a humidified atmosphere of 5% CO_2.
4) Forty-eight hours after initiation of enrichment, aspirate off the spent enrichment medium from the wells of the six-well plate and add 2 mL fresh, prewarmed enrichment medium.

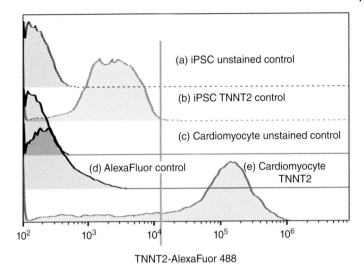

(a) iPSC unstained control

(b) iPSC TNNT2 control

(c) Cardiomyocyte unstained control

(d) AlexaFluor control

(e) Cardiomyocyte TNNT2

TNNT2-AlexaFuor 488

Figure 5.7 Flow cytometry data quantifying the level of TNNT2 expression in differentiated cardiomyocytes (e) in comparison to basal expression levels in the parental iPSCs (b) stained for TNNT2. Unstained populations for the parental iPSCs (a), cardiomyocytes (c) and a secondary antibody-stained control (d) all serve as negative controls for the analysis.

5) Return the cultures to the 37 °C incubator with a humidified atmosphere of 5% CO_2.
6) Repeat feeding of enrichment media every 48 hours until Day 7.
7) On Day 7, the enriched cardiomyocytes are ready to be analyzed via flow cytometry or used for further downstream applications (Figure 5.7).

5.4.3.3 Flow Analysis of PSC-Derived Cardiomyocytes

Pluripotent stem cell-derived cardiomyocytes are visualized via immuno-cytochemical staining of adherent samples or quantified using intracellular staining of singularized fixed cells using flow cytometry. This method describes intracellular staining of TNNT2 follow by flow cytometric analysis.

Note: this protocol is meant for harvesting cardiomyocytes from a single well of a six-well plate. If there is concern about the pellet size of cells being harvested, perform the protocol in a 2 mL microcentrifuge tube and adjust the volumes accordingly to scale.

Note: use the parental PSC as a negative control and differentiated cardiomyocytes as the sample.

1) Aspirate off the normal growth medium from the wells to be harvested.
2) Rinse the adherent cells with 2 mL of D-PBS (-/-) per well in a six-well dish.
3) Aspirate off the D-PBS wash, and add 1 ml TrypLE per well. Incubate the samples in a 37 °C, 5% CO_2 incubator for 5 minutes.
4) Triturate cells the cells 5–10 times with a P-1000 to generate a homogenous single cell suspension.
5) Transfer the cell suspension into a 2 mL microcentrifuge tube. Add 1 mL of RP20 Buffer to the well to collect any residual cells and pool with the original cell suspension. Mix well and take a small (50 μL) aliquot for cell counts.
6) Centrifuge the tubes at 200× g for 5 minutes.
7) Aspirate off the supernatant, add 1 mL 2% paraformaldehyde (PFA). Resuspend the cell pellet with a P-1000 to generate a homogenous single cell suspension.
8) Allow the cells to incubate in the fixation solution, at room temperature for 15 minutes.
9) Centrifuge cells at 200× g for 5 minutes at room temperature and aspirate off the supernatant without disturbing the cell pellet.
10) Resuspend the fixed cells in 1 mL of cold 90% methanol, per tube.
11) Incubate mixture at 4 °C for 15 minutes (over ice).
12) Following the 15 minute incubation, add 1 mL Flow Buffer to each tube and mix.
13) Centrifuge the cells at 200× g for 10 minutes at room temperature.
14) Aspirate off the supernatant and resuspend the cell pellet in 1 mL Flow Buffer per tube. Note: overnight storage of sample is not recommended since it results in increased background.
15) Transfer 0.5 to 1.0 × 10^6 cells into labeled microcentrifuge tubes filled with 1 mL Flow Buffer. Make sure to set up tubes for unstained controls.
16) Centrifuge the cells at 200× g for 5 minutes at room temperature.

17) Aspirate off the supernatant and resuspend the cell pellets in 1 mL of Flow Buffer per tube. Use a dilution of 1:2000 (V/V) of TNNT2 primary antibody; for staining controls, resuspend only in the Flow Buffer.

18) Incubate the cell suspensions for 1 hour at room temperature.

19) Centrifuge the cells at 200× g for 5 minutes at room temperature.

20) Aspirate off the solution and resuspend the cell pellets in 2 mL of Flow Buffer to gently wash the cells.

21) Centrifuge the cells at 200× g for 5 minutes at room temperature.

22) Aspirate off the solution and resuspend the cell pellets in 1 mL of Flow Buffer containing a 1:2000 (v/v) dilution of AlexaFluor 488 conjugated, goat anti-mouse IgG1 secondary antibody; for negative control, resuspend first tube with FB only – this tube is your unstained cells only control.

23) Incubate the mixture for 30 minutes at room temperature, protected from light.

24) Centrifuge the cells at 200× g for 5 minutes at room temperature.

25) Aspirate off the solution and resuspend the cell pellets in 2 mL of Flow Buffer per tube.

26) Centrifuge the cells at 200× g for 5 minutes at room temperature.

27) Aspirate off the solution and resuspend the cell pellets in 2 mL of Flow Buffer per tube.

28) Centrifuge the cells at 200× g for 5 minutes at room temperature.

29) Aspirate off the supernatant and resuspend the cell pellets in 500–µL of Flow Buffer.

30) Stain through caps of the FACS tubes and perform flow cytometry.

References

1 L.A. Williams, B.N. Davis-Dusenbery, K.C. Eggan. SnapShot: directed differentiation of pluripotent stem cells. *Cell* **149**, 1174–1174 e1 (2012).

2 S.N. Irion, M.C. Nostro, S.J. Kattman, G.M. Keller. Directed differentiation of pluripotent stem cells: from developmental biology to therapeutic applications. *Cold Spring Harbor Symposia on Quantitative Biology*, **LXXXIII**, 101–110 (2008).

3 Y. Yan *et al.* Efficient and rapid derivation of primitive neural stem cells and generation of brain subtype neurons from human pluripotent stem cells. *Stem Cells Transl Med* **2**, 862–870 (2013).

4 S. Tohyama *et al.* Distinct metabolic flow enables large-scale purification of mouse and human pluripotent stem cell-derived cardiomyocytes. *Cell Stem Cell* **12**, 127–137 (2013).

Index

Note: Pages references in *italics* refer to Figures; those in **bold** refer to Tables

Human Pluripotent Stem Cells: A Practical Guide, First Edition. Uma Lakshmipathy,
Chad C. MacArthur, Mahalakshmi Sridharan and Rene H. Quintanilla.
© 2018 John Wiley & Sons, Inc. Published 2018 by John Wiley & Sons, Inc.